Survival in the Cold
Hibernation and Other Adaptations

Proceedings of the International Symposium for Survival in the Cold
held in Prague, Czechoslovakia, July 2–5, 1980.

Editors:

X. J. Musacchia
Professor of Physiology and Biophysics
Dean of the Graduate School, University of Louisville, Louisville, Kentucky, U.S.A.

L. Jansky
Professor and Chairman
Department of Comparative Physiology, Charles University, Prague, Czechoslovakia

ELSEVIER/NORTH-HOLLAND
NEW YORK · AMSTERDAM · OXFORD

Published by:

Elsevier North Holland, Inc.
52 Vanderbilt Avenue, New York, New York 10017

Sole distributors outside USA and Canada:

Elsevier/North-Holland Biomedical Press
P.O. Box 211, Amsterdam, The Netherlands

Library of Congress Cataloging in Publication Data:

International Symposium for Survival in the Cold (1980:
    Prague, Czechoslovakia)
    Survival in the cold: hibernation and other adaptations.

    Includes bibliographical references and index.
    Contents: Seasonal aspects of reproduction in a hibernating rodent / Russell J. Reiter—
    Humoral control of hibernation in golden hamsters / L. Jansky . . . [et al.—Role of the
    endocrine glands in hibernation with special reference to the thyroid gland / Jack W.
    Hudson—[etc.] 1. Hibernation—Congresses. 2. Cold adaptation—Congresses.
    I. Musacchia, X. J., 1923- . II. Jansky, Ladislav. III. Title.
QL755.I48   1980   599.054'3   81-5482
ISBN 0-444-00635-4   AACR2

Manufactured in the United States of America

Contents

Preface

This text consists of the thematic lectures presented at the International Symposium "Survival in Cold" in Prague in July 1980. The symposium was oriented toward hibernation, cold adaptation, and thermogenesis. The programs and participants continued the tradition of meetings held previously in Boston (1959), Helsinki (1962), Toronto (1965), Aspen (1971), and Jasper (1977) and three other international meetings on related topics which were convened in Washington in 1968 and Prague in 1970 and 1974. In the intervening years, research interests in the biology of hibernation have moved from a base of qualitative observations into the realm of highly controlled experimental sciences.

The topics discussed at this symposium concerned mainly the humoral and neuronal mechanisms controlling physiological stimulation and inhibition mechanisms of heat production. A wide scope of problems was reviewed in seven sessions on: 1. biological rhythms and hibernation, 2. hormones and hibernation, 3. mechanisms controlling hibernation and sleep, 4. thermogenesis in hibernation and cold adaptation, 5. cellular functions in cold, 6. antimetabolic substances related to induction of hibernation, and 7. regulation of brown adipose tissue metabolism. The chapters in this text focus on recent advances in each of these subjects and areas for future experimentation.

It is evident that hibernation is a complex process which requires adaptation or modification at all levels of physiological organization. One of the most important aspects of the physiology of hibernation is the role of the endocrine glands. It has been a long standing assumption that endocrine inactivity is a necessary prerequisite for hibernation. Recent research in this area, however, indicates that hibernation is obviously under multiple hormonal control and that the older view can no longer be considered tenable. Although hibernation is a seasonal phenomenon, the hormonal systems of a hibernator undergo a number of rather divergent changes throughout the year. Seasonal changes in endocrines of hibernators and their role in physiological processes underlying hibernation have been investigated with many diverse experimental approaches, induced alterations in the endocrine milieu being among the most frequent.

The problem of photoperiodicity, the pineal gland, and reproduction in hibernating rodents is reviewed by R. J. Reiter. He shows that the pineal is absolutely necessary for synchronizing phases of the animal reproductive cycle with the appropriate season of the year and is essential for survival of the species including successful hibernation in the winter season. The importance of the pineal is also stressed in the review by L. Janský, Z. Kahlerová, J. Nedoma, and J. F. Andrews, who further illustrate the role of individual endocrine glands for induction of hibernation and show the important positive role of hormones such as melatonin, adrenal steroids, serotonin, and especially reverse triiodothyronine. On the other hand, it is shown that testosterone has a strong negative effect upon hibernating ability. Many studies indicate that the activity of the thyroid gland is reduced in animals several months prior to the onset of hibernation for reasons so far not clear.

The role of the thyroid gland and its relationship to hibernation examined in considerable detail in the chapter by J. W. Hudson. As clearly shown, not all species of hibernators require thyroid secretory inactivity as a pre-requisite for hibernation. There is an obvious difference between ground squirrels and hamsters in this respect. One question discussed is which aspect, if any, of the thyroid function should be considered as a general physiological adaptation of this gland in hibernation. It is emphasized that the endocrine systems of different species differ and that before any generalization is made the study of different species of hibernators is required. As indicated, reduced functional activity of the thyroid gland in some hibernators is important for the facilitation of fat deposition. The ground squirrel, for example, prepares for winter torpor by increasing the body fat deposits and by hypertrophy of brown adipose tissue. These modifications provide the primary energy sources during the hibernating season. On the other hand, such species as the hamster store considerable quantities of food, showing little weight gain in the fall and even some loss of body weight just prior to hibernation.

Precise knowledge of biochemical change in carbohydrate metabolism and regulatory mechanisms via gluconeogenic endocrine action is becoming critical. Various aspects of the problem of carbohydrate metabolism and the role of gluco-corticoids in the biology of hibernation are reviewed by X. J. Musacchia and D. R. Deavers. They show that some hibernators such as the ground squirrel utilize carbohydrates and do not maintain blood glucose levels during a period of hibernation; others such as hamsters show signs of regulated carbohydrate metabolism in which blood glucose levels are maintained during periods of hibernation. A role for glucocorticoids in hibernation is becoming evident. As demonstrated, a promising approach to the investigation of the functional role of carbohydrate metabolism in hibernation is the experimental model of helium-cold hypothermia in the hamster.

Recent concepts of thermogenesis are reviewed by L. C. H. Wang and B. Abbott. They report on factors which limit the magnitude and seasonal changes of maximum metabolism. These indicate that maximum metabolism can be predicted by body size in a manner comparable to that of basal metabolic rate. The results in two species of ground squirrels suggest that the maximum metabolism of a hibernator is relatively constant regardless of seasonal acclimatization, temperature acclimation, and hibernating status. This could be a result of the constituitive presence in the hibernator of inducible processes known to enhance maximum metabolism in nonhibernators. The chapter by B. Cannon, J. Nedergaard and U. Sundin is a milestone in the long-standing efforts of biochemists to understand mechanisms underlying thermogenic ability of brown adipose tissue. They focus on the presence of a specific protein in the mitochondria of brown adipose tissue for which the authors introduce the name "thermogenin." The amount of thermogenin and the capacity of brown adipose tissue for nonshivering thermogenesis are closely correlated. Since the levels of thermogenin are increased by acyl-CoA, it is suggested that acyl-CoA might be the physiological effector of thermogenesis in brown adipose tissue.

Considering the subject of hormonal thermogenesis it is evident that in addition to norepinephrine stimulated thermogenesis some other humoral sub-stances exert thermogenic action. Epinephrine, as an example, seems to act during the initial phase of cold stress, i.e., before the brown adipose tissue is fully developed and available for norepinephrine stimulated heat production. On the other hand, some other substances, such as reverse triiodothyronine (rT_3), a metabolite of thyroxine, might inhibit resting and epinephrine stimu-lated metabolism by acting directly in target organs. It is provacative to consider that the resting metabolism of hibernators reveals a component of nonshivering thermogenesis which could be safely suppressed to induce hiberna-tion. From this point of view, the studies of physiological inhibition of processes contributing to that component of nonshivering thermogenesis are promising.

The first suggestions that some controlling hormone might be responsible for the induction of hibernation were made more than 40 years ago. The present state of our knowledge of relationships of antimetabolic agents to the induction of hibernation is reviewed by H. Swan. Contrary to the old view, it is no longer evident that brown adipose tissue is an endocrine organ involved in the induction of hibernation. On the other hand, there appear to be a number of substances which are able to suppress the metabolic rate and to lower the body temperature of the recipient. Some of these include antabolone isolated from the brain of aestivating fish, neurotensin, bombesin, and the substance isolated from the blood of some hibernators called "hibernation induction trigger" (HIT). The latter substance has been repeatedly tested from the physiological point of view. P. Oeltgen and W. Spurrier report on chemical characterization and specific isolation of the HIT molecule. They present recent progress in the development of a rapid and specific in vitro and in vivo assay, including utilizing nonhibernators such as monkeys. Their chapter focuses on the role of HIT, its biochemical role in the physiological phenomenon of hibernation and some potential clinical aspects of the HIT molecule.

H. C. Heller, T. S. Kilduff and F. R. Sharp report on a relatively new approach to provide a more global insight into brain functions. Experiments are described in which (^{14}C) deoxyglucose was injected via chronically implanted jugular catheters in animals in different stages of hibernation or during euthermia with thermogenesis driven by cooling of the skin or hypothalamus. They show clear differentation between animals in forced hypothermia and in hibernation. Hibernating animals remain responsive to sensory stimuli and brain sensory structures show high levels of glucose utilization.

The integrity of the central nervous system, which ensures survival at low body temperature, is based on altered properties of neuronal membranes, especially in the synaptic region. As H. Rahmann and R. Hilbig report, the brain gangliosides containing sialic acid residues are involved in the processes of membrane temperature adaptation. Correlations are demonstrated among systematic position of species, thermal adaptation status, and brain ganglioside composition. A decrease of the environmental temperature induces a long-term formation of more polar polysialogangliosides. These adaptive changes are credited with maintaining the polarity of neuronal membranes under new thermal conditions. Much of the present knowledge of the neural basis of hibernation is largely the result of EEG recordings from a limited number of brain sites.

Adaptation of cellular metabolism to low body temperatures is mainly brought about by the increased affinity of enzymes for substrates as documented in the review by H. W. Behrisch, D. H. Smullin and G. A. Morse. This research indicates that the thermal change rather than the low temperature itself is a major environmental disturbance to an organism exposed to cold. Physiological and biochemical studies on hibernators suggest that the animal exists in different biochemical states at different times of the year. It is emphasized that metabolic reorganization in a hibernator occurs in a complex pattern in which certain tissues and metabolic sequences take temporary precedence.

Many of these researches show that hibernation as a biological phenomenon secondary to homeothermy represents a complex series of regulatory mechanisms which are common to hibernators as well as other mammals. A comprehensive understanding of hibernation phenomena must focus on the physiological and biochemical events which are specifically regulated, the nature of such specificity, and finally, the potential for induction of hibernating torpor and how to imitate it in other mammals. Dealing with this widely divergent area of biological research, investigators at the present time are focusing their efforts mainly on endocrine functions and to a lesser extent on the area of the integrative and regulatory role of the nervous system. Results, however, show clearly that we lack information about the regulatory role of the autonomic nervous system and about neurotrophic relations. Neurotrophic functions, especially, can be considered an important process having integrative and regulatory roles. This area of research seems to be very promising for biologists interested in hibernation since the time course of neurotrophic related changes corresponds well to the seasonal changes in a hibernator.

Substantial differences in adaptation of elementary cell functions during hibernation strongly indicate the significance of integrative and regulatory systems during hibernation. Without more precise knowledge of the regulation of integrative systems, it will not be possible to make any generalizations concerning hibernation and apply the results obtained to other mammals, including man. From the view of possible application, the experiments with the HIT molecule seem to be most promising.

Finally, it should be known that in addition to this text, abstracts of 71 papers have been published in *Cryobiology*, Volume 18, 1981 and 44 of these voluntary papers will be published in short text in Acta Universitatis Carolinae, Ser. Biologica, 1981, Charles University, Prague.

Jan Moravec
Ladislav Janský
X. J. Musacchia

Acknowledgments

This text was developed and completed because of the cooperation and efforts of many researchers and several institutions: the Czechoslovak Medical Society J. E. Purkyne, the Czechoslovak Physiological Society, Charles University, the University of Louisville, the Society for Cryobiology, and the International Hibernation Society, under the supervision of the Commission for Environmental Physiology at the International Union of Physiological Sciences. In particular, we want to express our sincere thanks to the staff personnel of the Department of Comparative Physiology, Charles University and the Graduate School of the University of Louisville, Kentucky, who provided conference support.

Special mention is made of other individuals who served on the Organizing Committee, Dr. J. Mejsnar, Dr. S. Vybíral and Dr. B. Štefl. The assistance and support provided by Jona Klimertová and many other friends and colleagues in Prague are also happily acknowledged.

The editors want to express their sincerest thanks to B. G. Ash, Sharon Mills and Diane Cardenas of the University of Louisville for their unstinting efforts in the preparation of the press manuscript: copy editing, typing, revising and proofing of the entire text.

The many agencies and foundations which contributed to specific research efforts were acknowledged by the authors in their individual chapters.

There are many who are not named, but who also contributed. To them, we extend our sincere thanks and our hope for their understanding that not all who contribute can be individually named.

Lastly, we are thankful for the large numbers of invited participants who contributed to this International Symposium and made it a splendid opportunity to exchange the newest information in our area of research.

X. J. Musacchia
L. Janský
Jan Moravec

Contributors

BRUCE ABBOTTS
Department of Zoology
University of Alberta
Edmonton, Alberta, Canada T6G 2E9

J. F. ANDREWS
Department of Physiology
Trinity College
Dublin, Ireland

H. W. BEHRISCH
Institute of Arctic Biology
University of Alaska
Fairbanks, Alaska 99701, U.S.A.

BARBARA CANNON
Wenner-Gren Institute
University of Stockholm
Norrtullsgatan 16
S-113 45, Stockholm, Sweden

D. R. DEAVERS
Physiology/Pharmacology Discipline
College of Osteopathic Medicine and Surgery
3200 Grand Avenue
Des Moines, Iowa 50312, U.S.A.

H. CRAIG HELLER
Department of Biological Sciences
Stanford University
Stanford, California 94305, U.S.A.

R. HILBIG
Institute of Zoology
University of Stuttgart-Hohenheim
7 Stuttgart 70 (Hohenheim), FRG

JACK W. HUDSON
Department of Biology
University of Alabama in Birmingham
University Station
Birmingham, Alabama 35294, U.S.A.

L. JANSKÝ
Department of Comparative Physiology
Faculty of Science
Charles University
128 44 Prague 2, Viničná 7
Czechoslovakia

Z. KAHLEROVÁ
Department of Comparative Physiology
Faculty of Science
Charles University
128 44 Prague 2, Viničná 7
Czechoslovakia

THOMAS S. KILDUFF
Department of Biological Sciences
Stanford University
Stanford, California 94305, U.S.A.

G. A. MORSE
Institute of Arctic Biology
University of Alaska
Fairbanks, Alaska 99701, U.S.A.

X. J. MUSACCHIA
Department of Physiology and Biophysics
University of Louisville
Louisville, Kentucky 40292, U.S.A.

JAN NEDERGARRD
Wenner-Gren Institute
University of Stockholm
Norrtullsgatan 16
S-113 45, Stockholm, Sweden

J. NEDOMA
Department of Comparative Physiology
Faculty of Science
Charles University
128 44 Prague 2, Viničná 7
Czechoslovakia

PETER R. OELTGEN
Department of Pathology
University of Kentucky
College of Medicine
Lexington, Kentucky 40511, U.S.A.

H. RAHMANN
Institute of Zoology
University of Stuttgart-Hohenheim
7 Stuttgart 70 (Hohenheim), FRG

RUSSEL J. REITER
Department of Anatomy
The University of Texas
Health Science Center at San Antonio
7703 Floyd Curl Drive
San Antonio, Texas 78284, U.S.A.

FRANK R. SHARP
Department of Neurosciences
Medical Center
University of California at San Diego
San Diego, California 92037, U.S.A.

D. H. SMULLIN
Institute of Arctic Biology
University of Alaska
Fairbanks, Alaska 99701, U.S.A.

WILMA A. SPURRIER
Division of Neurological Surgery
Loyola University of Chicago
Stritch School of Medicine
2160 South First Avenue
Maywood, Illinois 60153, U.S.A.

ULF SUNDIN
Wenner-Gren Institute
University of Stockholm
Norrtullsgatan 16
S-113 45, Stockholm, Sweden

HENRY SWAN
Department of Clinical Sciences
Colorado State University
Ft. Collins, Colorado 80521, U.S.A.

LAWRENCE C. H. WANG
Department of Zoology
University of Alberta
Edmonton, Alberta, Canada T6G2E9

Participants

ALEXSON, Stefan, Wenner-Gren Institute, Norrtullsgatan 16 S-11345 Stockholm,
    Sweden
AMBID, Louis, Institute of Physiology, Rue F. Magendie, 31400 Toulouse, France
ANDREWS, James F., Department of Physiology, Trinity College, Dublin 2, Ireland
ACQUIN, Patricia, 4200 East 9th Ave., Denver, Colorado 80262, USA
AUGUSTINOWICZ, Stanislaw, Al. Revolucji Pazdziernikowej 62, 01-424 Warszawa,
    Poland
BANCHERO, Natalio, 4200 E. 9th Ave., Denver, Colorado 80262, USA
BASS, Arnost, Institute of Physiology, Czechoslovak Acad. Sci., Vídeňská 1083,
    142 20 Prague-Krč, ČSSR
BAZHENOV, Yu.I., Institute of Physiology, AN Kirghiz SSR, Frunze, USSR
BAZHENOVA, A.F., Institute of Physiology, AN Kirghiz SSR, Frunze, USSR
BEHRISCH, H., Institute of Arctic Biology, University of Alaska, Fairbanks,
    Alaska, 99701, USA
BOŠTÍK, Josef, Department of Comparative Physiology, Charles University,
    Vinična 7, 12844 Prague 2, ČSSR
CANNON, Barbara, Wenner-Gren Institute, Norrtullsgatan 16 S-11345 Stockholm,
    Sweden
CORNEHEIM, Claus, Wenner-Gren Institute, Norrtullsgatan 16 S-11345 Stockholm,
    Sweden
CRANFORD, Jack A., Department of Biology, Virginia Polytechnic Institute & State
    University, Blackburg, Virginia 24061, USA
DETERING, Ricarda, Physiological Institute I, Nussallee 1, 5300 Bonn 1, FRG
DRAHOTA, Zdenek, Institute of Physiology, Czechoslovak Acad. Sci., Vídeňská
    1083, 142 20 Prague-Krč, ČSSR
DVOŘÁK, Richard, Institute of Experimental Medicine, Czechoslovak Acad. Sci.,
    Legerova 61, Prague 2, ČSSR
FLORANT, Gregory L., Department of Neurology, Montefiore Hospital, 111 E. 210th.
    St., Bronx, New York 10467, USA
HALBERG, Franz, Chronobiology Lab., 5-187 Lyon Lab., 420 Washington Ave.,
    Minneapolis, Minnesota 55455, USA
HANUŠ, Karel, IKEM, Videnska 800, 146 22 Prague 4-Krč, ČSSR
HARTNER, William C., Northwestern University, Dept. of Biology, 360 Huntington
    Ave., Boston, Massachusetts 02115, USA
HAWKINS, Mike, 2217 West Elizabeth, Apt. 303, Fort Collins, Colorado 80521, USA
HELDMAIER, Gerhard, Zoological Institute, Siesmayerstr. 70, 6000 Frankfurt/M.,
    FRG
HELLER H., Craig, Department of Biological Sciences, Stanford University,
    Stanford, California 94305, USA
HILBIG, Reinhard, Institute of Zoology, 7000 Stuttgart 70 (Hohenheim) FRG
HOLEČKOVA, Emma, Institute of Physiology, Czechoslovak Acad. Sci., Vídeňská
    1083, 142 20 Prague -Krč, ČSSR
HOUŠTĚK, Josef, Institute of Physiology, Czechoslovak Acad. Sci., Vídeňská 1083,
    142 20 Prague -Krč, ČSSR
HUBL, Ludger, Physiological Institute I, Nussallee 11 5300 Bonn 1, FRG
HUDSON, Jack W., University Station, Birmingham, Alabama 35294, USA
HULBERT, Anthony John, Department of Biology, University of Wollongong,
    Wollongong 2500, Australia
ILLNEROVA, Helena, Institute of Physiology, Czechoslovak Acad. Sci. Vídeňská
    1083, 142 20 Prague -Krč, ČSSR

xii

ISABAYEVA, V. V., Institute of Physiology, Prospekt Mira 56 720044 Frunze, Kirg.
USSR
JANSKÝ, Ladislav, Department of Comparative Physiology, Charles University,
Viničná 7, 128 44 Prague 2, ČSSR
KAHLEROVA, Zuzana, Department of Comparative Physiology, Charles University,
Viničná 7, 128 44 Prague 2, ČSSR
KAIGL, Ctibor, 1901 E. Thomas Blvd., Phoenix, Arizona 85016, USA
KOLAEVA, Stella G., Institute of Biological Physics, Acad. Sci. USSR, Pushchino,
Moscow Region 142292, USSR
KOLÁŘ, František, Department of Comparative Physiology, Charles University,
Viničná 7, 128 44 Prague 2, ČSSR
KOPECKY, Jan, Institute of Physiology, Czechoslovak Acad. Sci., Vídeňská 1083,
142 20 Prague -Krč, ČSSR
KUČERA, Vladimír, Cardiocenter, Faculty Hospital Motol, V úvalu 84 150 06 Prague
5, ČSSR
LECHNER, Andrew, 4200 East 9th Ave., Denver, Colorado 80262, USA
LEMONS, Daniel E., Box 38, 630 W. 168th Street, New York, N.Y. 10032, USA
LYNCH, Marina, Department of Physiology, Trinity College, Dublin, Ireland
MALAN, Andre, CNRS, 23 rue Becquerel, F-67087 Strasbourg C, France
MEJSNAR, Jiří, Department of Comparative Physiology, Charles University, Viničná
7, 128 44 Prague 2, ČSSR
MISCHER, Ortwin, Abt. Pathobiochemie, Med. Akademie Magdeburg, Leipziger Str.
44, 301 Magdeburg, GDR
MOHELL, Nina A., Wenner-Gren Institute, Norrtullsgatan 16, S-11345 Stockholm,
Sweden
MOORE, Gregory A., Wenner-Gren Institute, Norrtullsgatan 16, S-11345 Stockholm,
Sweden
MORAVEC, Jan, Department of Comparative Physiology, Charles University, Viničná
7, 128 44 Prague 2, ČSSR
MORIYA, Kiyoshi, Faculty of Education, Hokkaido University Sapporo, Japan
MUSACCHIA, X.J., Graduate School, Belknap Campus, University of Louisville,
Kentucky 40292, USA
NANBURG, Eewa, Wenner-Gren Institute, Norrtullsgatan 16 S-11345 Stockholm,
Sweden
NECHAD, Myriam, Wenner-Gren Institute, Norrtullsgatan 16 S-11345 Stockholm,
Sweden
NEDERGAARD, Jan, Wenner-Gren Institute, Norrtullsgatan 16 S-11345 Stockholm,
Sweden
OELTGEN, Peter R., Department of Pathology, University of Kentucky, College of
Medicine, Lexington, Kentucky 40511, USA
PAJUNEN, Irmeli, Arkadiank 7, SF 00100 Helsinki 10, Finland
PARKER, Glenn, Biology Department, Laurentian University Sudbury, Ontario,
Canada P3E 2C6
PASTUKHOV, Yury, Institute of Biological Problems of the North, Far-East Sci.
Center, Magadan, K. Marks Str. 24, USSR
PEHOWICH, Daniel J., Department of Zoology, University of Alberta, Edmonton,
Alberta, Canada T6G 2E9
PETROVIČ, Vojislav M., Faculty of Sciences, Studentski trg 16, Beograd,
Yugoslavia
PICHOTKA, J., Institute of Physiology, University of Bonn Nussallee 11, 53 Bonn,
FRG
POGZOPKO, Piotr, Institute of Animal Physiology, 05110 Jablonna near Warsaw,
Poland
POHL, Hermann, Max Planck Institute for Physiology of Behaviour, 8131 Andecns,
FRG
POPOVA, Nina K., 630090 Novosibirsk 90, Institute of Cytology and Genetics, USSR

RAFAEL, Johannes, Institut für Biochemie I, Im Neuheimer Feld 328, D-6900
  Heidelberg, FRG
REINHARD, Friedrich Gustave, Max Planck Institut für Biochemie, Am Klopferspitz
  8033 Martinsried, FRG
REITER, Russel, 7703 Floyd Curl Drive, San Antonio, Texas 78284, USA
RESCH, Garth, University of Missouri, Department of Biology 5100 Rockhill Road,
  Kansas City, Missouri 64110, USA
SCHATTE, Christopher, Colorado State University, Department of Physiology, Fort
  Collins, Fort Collins 80524, USA
SCOTT WORKMAN, Grace, 17 Braemar Ave., Toronto, Ontario, Canada M5P 2LK
SCOTT, Irena, Department of Physiology, Ohio State University, 310 Hamilton
  Hall, Columbus, Ohio 54210, USA
SIGMAN, Harvey H., Jewish General Hospital, 3755 Cote St. Catherine Rd.,
  Montreal, Quebec, Canada H3T IE2
SIMMONDS, Richard Carrol, 4301 Jones Bridge Road, Bethesda, Maryland 20014, USA
SOCHACKA, Zofia, Al. Revolucji Pazdiernikowej 62, Warszawa 01424, Poland
SPURRIER, Wilma A., Division of Neurological Surgery, Loyola University of
  Chicago Stritch School of Medicine, 2160 S. First Ave., Maywood, Illinois
  60153, USA
STEINLECHNER, Stephan, Zoological Institute, Siesmayerstr. 70, 6000 Frankfurt/
  Main, FRG
STRUNECKÁ, Anna, Department of General Physiology, Charles University, Viničná
  7, 12844 Prague 2, ČSSR
SUNDIN, Ulf, Wenner Gren Institute, Norrtullsgatan 16 S-11345 Stockholm, Sweden
SVARTENGREN, Jan, Wenner Gren Institute, Norrtulsgatan 16 S-11345 Stockholm,
  Sweden
SVOBODA, Petr, Institute of Physiology, Czechoslovak Acad. Sci., Vídeňská 1083,
  142 20 Prague-Krč, ČSSR
SWAN, Henry, 6700 W. Lakeridge Road, Lakewood, Colorado 80227, USA
ŠTEFL, Bohumir, Department of Comparative Physiology, Charles University,
  Viničná 7, 128 44 Prague 2, ČSSR
TÄHTI, Hanna, Department of Biomedical Sciences, Box 607, SF 33101 Tampere 10,
  Finland
TOTH, Delphi M., University of Oklahoma, College of Medicine Oklahoma City,
  Oklahoma 73190, USA
TWENTE, John, University of Missouri-Columbia, Dalton Research Center, Research
  Park, Columbia, Missouri 65201, USA
TWENTE, Janet, University of Missouri-Columbia, Dalton Research Center, Research
  Park, Columbia, Missouri 65201, USA
VÁRADY, Jozef, University of Physiology of Farm Animals, Slovak, Acad. Sci.,
  Palackého 12, 040 00 Košice, ČSSR
VYBÍRAL, Stanislav, Department of Comparative Physiology, Charles University,
  Viničná 7, 128 44 Prague 2, ČSSR
VYSKOČIL, Frantisek, Institute of Physiology, Czechoslovak Acad. Sci., Vídeňská
  1083, 142 20 Prague 4 Krč, ČSSR
WALKER, James H., University of California, Div. of Nat. Sciences, Santa Cruz,
  California 95064, USA
WALTON, J. Bernard, Department of Physiology, Trinity College, Dublin 2, Ireland
WANG, Lawrence C.H., Department of Zoology, University of Alberta Edmonton,
  Alberta, T6G 2ED Canada
WERNER, Roderich, Institute of Zoology, Department of Zoophysiology, University
  of Kiel, Olshausens. 40-60, 2300 Kiel, FRG
WUNDER, Bruce Arnold, Institute of Zoology, Siesmayerstr. 70, D 6000
  Frankfurt/M., FRG
Wünnenberg, Wolf, Institute of Zoology, University of Kiel, Ohlshausenstr.
  40-60, 2300 Kiel, FRG

SEASONAL ASPECTS OF REPRODUCTION IN A HIBERNATING RODENT:
PHOTOPERIODIC AND PINEAL EFFECTS

RUSSEL J. REITER
Department of Anatomy, The University of Texas, Health Science Center at San
Antonio, 7703 Floyd Curl Drive, San Antonio, Texas 78284, U.S.A.

INTRODUCTION

A number of mammals circumvent the cold days of winter by undergoing hibernation during a portion of each year. Hibernation is a complex process which
requires adaptations or modifications at all levels of physiological organization. The changes associated with the hibernating process have been investigated with many diverse experimental approaches [1, 9, 14, 19, 20, 31], yet
there is no consensus as to the absolute prerequisites for hibernation. Induced
alterations in the endocrine milieu are among the approaches that have been
rather widely exploited in an attempt to define the physiological processes underlying hibernation [4, 13, 15, 16]. Here again, uniform experimental findings
and unanimity of opinion are often lacking.

The involvement of the reproductive system in the hibernating process has
long been a subject of intense interest [15, 37]. In most hibernators, reproductive capability often fluctuates on a seasonal basis not unlike the manner in
which thermoregulatory processes change. Thus, during hibernation the reproductive organs of both males and females are frequently atrophic. The precise
relationships of the hibernating and the reproductive cycles vary among species
[8, 10, 17]. In the species of interest in the present report, i.e., the Syrian
hamster (*Mesocricetus auratus*), it has been reported that successful prolonged
hibernation can occur only in individuals in which the gonads are atrophic [29,
30]. Hence, perturbations which prevent gonadal involution also interfere with
the thermoregulatory responses required for hibernation.

One technique which reproducibly prevents gonadal atrophy in both male and
female Syrian hamsters during the winter months is pinealectomy [3, 24]. Even
before these observations were made, Mogler [20] claimed that the activity of
the pineal gland and the activity of the reproductive system were inversely
related in the Syrian hamster. Thus, if dormant reproductive organs are required for successful hibernation in this species, as has been proposed [29,
30], then surgical removal of the pineal gland undoubtedly would interfere with
the thermoregulatory responses necessary for winter torpor. The discussion

which follows is concerned with the photoperiodic regulation of the annual reproductive cycle in the Syrian hamster and its potential relationships to hibernation.

RESULTS AND DISCUSSION

The annual reproductive cycle--relation to hibernation: Most nonhuman mammals that normally reside in the temperate and polar regions, where annual fluctuations in the environment are substantial, breed on a seasonal basis. These seasonal reproductive cycles are under the influence of a variety of factors, one of the most important of which is day length. It is essential that the circannual reproductive rhythms be precisely timed if the species is to survive in an ever changing environment. The important event of the annual cycle is the birth of the young which must occur in the spring and early summer, thus providing the newborns with maximum chance for survival.

Many animals respond to the changing photoperiod by making proper physiological adjustments to prepare for the forthcoming seasons. In an anthropomorphic sense, it is not only necessary for animals to be aware of the season at hand, but they must also be able to predict what the environmental situation will be months in the future. Because of the regularity with which the photoperiod changes, it has great value for species whose existence depends on carefully regulated reproductive cycles. Since it is annually reproducible, animals have evolved mechanisms whereby the photoperiodic cycle has distinct advantages. The organ that adjusts the reproductive status of these animals in concert with the appropriate season is the pineal gland.

The pineal gland responds to changes in day length by producing hormones, primarily inhibitory substances, which determine the animals' sexual capability [18, 28]. Generally, long days are considered to be inhibitory to the pineal gland and consequently stimulatory to reproduction. Short days, because they augment the biochemical and endocrine activity of the pineal, suppress reproductive function. The reproductive potential of long-day breeding rodents, such as the Syrian hamster, are readily manipulated by the photoperiod with the pineal serving as the essential intermediary between the photic environment and the neuroendocrine-reproductive axis. In the absence of the pineal gland, cyclic changes in reproduction are lost because the animal is incapable of responding to the photoperiod [3, 24]. After pinealectomy animals become continuous rather than seasonal breeders.

The generalized scheme, summarized in Fig. 1, describes the provisional interactions of the photoperiod with the pineal gland and reproduction in long-

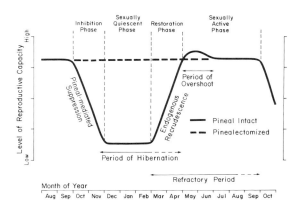

Fig. 1. Schematic representation of the proposed relationships of the photo-period with the annual reproductive cycle and hibernation period of the Syrian hamster. Phases of the annual cycle are given at the top of the figure. In the fall of the year, the pineal produces inhibitory factors, and induces atrophy of reproductive organs. This allows the animals to enter prolonged hibernation. As spring approaches, but before animals emerge from hibernation, the neuro-endocrine axis becomes refractory to the inhibitory influence of the pineal gland and, as a consequence, the reproductive organs recrudesce. The period of gonadal regrowth is associated with less frequent hibernation bouts and event-ually with terminal arousal. The animals reproduce very early in the sexually active phase, especially during the period of overshoot. Pinealectomized hamsters never experience the seasonal waxing and waning of reproductive cap-ability.

day breeding rodents. The data on which these theoretical relationships are based were derived primarily from experiments in which the Syrian hamster was the experimental model. However, the model presumably has applicability to other species as well, especially to other long-day breeding hibernating rodents [25]. For the purpose of the description that follows, the annual reproductive cycle has been divided into four phases: the inhibition phase, the sexually quiescent phase, the restoration phase, and the sexually active phase [26].

Inhibition phase: During the long days of the summer, hamsters are repro-ductively active and competent. As fall approaches, however, daylight falls below a certain critical length, which for the hamster is 12.5 hours of light

[5, 6], and the antigonadotrophic activity of the pineal initiates gonadal involution. This identifies the onset of the inhibition phase of the cycle (Fig. 1). Since the hamster is a fossorial hibernating species, it is believed that behavioral changes also influence the reproductive state. Presumably, in the late summer and fall hamsters spend increased amounts of time in their subterranean burrows preparing for hibernation. This further restricts the amount of light to which they are exposed and provides the pineal with an additional stimulus to produce and secrete its antigonadotrophic compounds. The net effect of this severe light restriction is pineal-induced gonadal atrophy.

During the inhibition phase, the reproductive organs atrophy in a matter of weeks. If animals are pinealectomized prior to their exposure to the shortening days of the fall, reproductive involution does not ensue [3, 24]. The regressive changes in the gonads are very dramatic. The testes become aspermatogenic and gonadal steroid production drops; pituitary and plasma levels of gonadotrophins, including luteinizing hormone (LH), follicle stimulating hormone (FSH) and prolactin, also exhibit significant decreases. In females, ovulation ceases, the animals become anestrus, uteri involute, and the hormonal levels are changed accordingly. Presumably, near the end of the inhibition phase hibernation is initiated (Fig. 1). At this time the status of the reproductive systems is compatible with the entrance of the animal into hibernation [12, 30].

Sexually quiescent phase: After the animals experience total reproductive suppression, they enter the sexually quiescent phase of their annual cycle. During this period the hamsters are endocrinologically and behaviorally incapable of sexual reproduction. Proof that the pineal gland is required to force gonadal regression during the inhibition phase and also maintain the gonads in an atrophic state during the sexually quiescent phase is easily provided by subjecting animals with atrophic gonads to pinealectomy. This results in nearly an immediate initiation of gonadal regrowth with full reproductive potential being achieved within about eight weeks [21]. The sexually quiescent phase of the cycle can also be prematurely interrupted by moving hamsters from the short days of the winter into the long days of the laboratory. This, again, is a consequence of the inhibition of the antigonadotrophic activity of the pineal by the long days. The sexually quiescent phase presumably is associated with the most successful hibernation period of the animals.

If, in fact, hamsters are pinealectomized before entering this phase, they remain sexually competent, breed, and the females deliver young [23]. However, the survival of the young during the cold days of the winter is negligible and virtually all are lost to the harsh environmental conditions within the first several days after birth.   Thus, it is obvious that having litters during the winter months would have little value in replenishing the species.   In addition, it can be argued that it would be detrimental to the species because of the wasted energy associated with the intrauterine growth and delivery of winter young with little chance for survival.   In essence, the short days of the winter tend to spare the animals from indiscriminate pregnancies which would be unproductive in terms of increasing the number of animals.   Secondarily, the reduced day lengths aid in promoting hibernation.

Restoration phase: As spring approaches the reproductive organs of both male and female hamsters begin to recrudesce [24].   Since the shorter days in the fall cause gonadal atrophy, the assumption was that the increasing day lengths of spring promote the regrowth of the gonads. However, there was suspicion that this was not the case because reproductive recrudescence begins during the later stages of hibernation when the animals are still in darkness. When examined, it was found that the restoration of normal reproductive function in hamsters is indeed a light independent phenomenon [22, 27].   Thus, even if totally deprived of light stimuli by blinding, the gonads eventually reach maturity.

The regrowth of the gonads is referred to as the restoration phase of the annual reproductive cycle (Fig. 1).   As noted above, it is important for hibernating species to begin restoration of reproductive function before they actually emerge from hibernation since this allows them to mate immediately after terminal arousal. If recrudescence was delayed until arousal, the animals would have to wait an additional period of time (roughly eight weeks) during which their gonads would regrow to the functional state.   This delay would effectively shorten the summer breeding season.

The regrowth of the gonads during the restoration phase was speculated to result from the increased sensitivity of the neuroendocrine-reproductive axis to the stimulatory effect of light.   This does not seem to be the mechanism, however, since recrudescence occurs without the benefit of light.

Rather, this investigator maintains that normal reproductive function is reestablished because the neuroendocrine-gonadal system becomes refractory to the suppressive effects of darkness on the pineal gland [22] (Fig. 1).   It is also known that in the later stages of hibernation both the frequency and the duration of the hibernating bouts decrease [8].   This may be a consequence of increasing levels of circulating gonadal steroids from the growing reproductive organs.   Under any circumstances, the net effect of the spontaneous recrudescence of the gonads results in animals being sexually competent immediately upon their emergence from hibernation.

Sexually active phase: Hamsters are likely to engage in sexual activity very soon after they are seen above ground in the early spring. Curiously, when the gonads recrudesce, they grow to a size which exceeds their normal laboratory weight (period of overshoot, Fig. 1). This period is associated with higher than normal levels of both pituitary and plasma gonadotrophins [28]. Although the phenomenon of overshoot has not been extensively studied, it has been noted in several experiments. It is during this phase that hamsters are most active sexually and, because of the short gestation in the female (roughly 16 days), the first litters are delivered within weeks after the termination of hibernation. It is likely that females of the species have several additional litters during the sexually active phase.

Since the onset of refractoriness initiates the restoration phase allowing the gonads to recrudesce, the question was asked, does the neuroendocrine-reproductive axis of these animals remain refractory during the sexually active phase? When subjected to test, it was confirmed that if animals in the early sexually active phase are deprived of light their gonads do not involute, i.e., they are indeed refractory to the stimulated pineal gland [22, 27].

The next question was, what role does light play in the annual reproductive cycle of the hibernating hamster? It is the absence of light which results in the pineal-mediated gonadal regression in the fall of the year. It was reasoned, also, that it is the darkness of the subterranean burrow which keeps the gonads in an atrophic condition during the winter hibernation period. Finally, during the restoration and sexually active phases, the gonadal system is refractory to the inhibitory influence of darkness acting by way of the pineal gland. Thus, the entire annual cycle could apparently proceed without any photoperiodic stimuli.

Some experimental data have accumulated which indicates this is not quite the case. During the sexually active phase some light must be available for a certain number of days to interrupt the refractory period. The specific duration of photoperiodic stimulation is not agreed upon but it seems to fall between 10 and 22 weeks under the experimental conditions in which it was examined [27, 32]. Interruption of the refractory period ensures that the short days of the subsequent fall will again induce gonadal atrophy.

A recent study illustrates the tenacious nature of the refractory period. The purpose of the study was to determine whether gonadal involution would in fact be prevented during the cold short days of the winter if the refractory

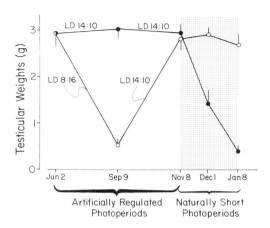

Fig. 2. The effect of refractoriness of the neuroendocrine-reproductive axis of male hamsters on gonadal involution induced by naturally short days during the winter months.   Hamsters whose gonads had undergone atrophy (due to exposure to short light:dark cycles, i.e., LD 8:16) and subsequent regeneration in long days (LD 14:10) became refractory to darkness.   As a consequence, exposure to the naturally short and cold days of winter was not followed by reproductive regression.

period was not interrupted. The results of the experiment are summarized in Fig. 2.

Beginning on June 2, laboratory maintained hamsters were placed in short days (light:dark cycles of 8:16) for a period of 14 weeks. By September 9, the gonads of these animals had become completely atrophic. The animals were then moved to long daily photoperiods (light:dark cycles of 14:10) for a period of eight weeks until November 8. This period of long days was associated with recrudescence of the reproductive organs and the development of refractoriness.

On November 8, these refractory animals, along with appropriate controls which were not in the refractory condition, were moved to the naturally short days of winter. As expected, the control animals experienced pronounced gonadal regression typical of animals in the nonrefractory condition.   Conversely, the reproductive organs of hamsters in the refractory condition remained sexually mature well into the winter months (Fig. 2).   These findings prove it is essential for the animals to interrupt the refractory period before experiencing a second winter.   If they fail to do this, their gonads would not involute in response to the short days and hibernation would likely be impaired.

Thus, the annual reproductive cycle of long-day breeding hibernating species is closely linked with the prevailing environmental photoperiod. It is hormones from the pineal gland which signal the reproductive system of the photoperiodic condition. One of the important pineal hormones in this scheme may be melatonin. When properly administered, it has been shown to duplicate the effects of photoperiod on reproduction in the hamster [28]. There are, however, other pineal compounds which should not be overlooked for their potential as gonad regulating substances [2, 35].

Until about two decades ago the pineal gland was considered by many to be a functionless vestige. It is obvious from recent experiments that in some species the gland is absolutely necessary to synchronize the phases of the annual reproductive cycle with the appropriate seasons of the year. This is essential for the survival of the species. Without the pineal gland, hamsters, it can be argued, would indiscriminately breed throughout the year and many of the young, because they would be born during environmentally harsh seasons, would not survive. This is a condition that animals in their natural habitat can not tolerate. Even with the birth of the young during spring, the optimal season, many of the newborns are lost to predation, flooding and other environmental factors.

How many species depend on the pineal gland to synchronize their annual reproductive cycle remains unknown. Certainly, these relationships apply in the Syrian hamster and to a number of other rodent species [11, 38]. However, recent findings about the importance of photoperiod stimuli in influencing reproduction suggest the interactions between the pineal gland and the gonads described above may be applicable to other orders of mammals as well [7, 33, 34]. Even in humans, the pineal secretory products may have yet undefined effects [36]. Undoubtedly, during the next decade research in this area will be intensive and discoveries will be widespread.

SUMMARY

Seasonal reproduction is a prerequisite for animals which inhabit temperate and polar regions of the earth where, during roughly half of each 12-month period, they experience cold environmental temperatures and shortened photoperiods. It is imperative that the young of these species are not born during the winter months inasmuch as their survival at this time of the year would be jeopardized by the harsh environmental temperatures and reduced food supply. A number of different mechanisms have evolved in mammals to guarantee seasonal reproduction.

In the Syrian hamster, a hibernating rodent, the reproductive organs involute as a consequence of the short days of the fall and winter. Thus, the testes of the males become aspermatogenic and the females become anovulatory. The regression of reproductive organs during the winter months serves a twofold purpose, namely, it prevents the animals from reproducing during this season and it ensures successful prolonged hibernation.

The pineal gland is known, because of its secretion of antigonadotrophic products, to be involved in the atrophic responses of the reproductive organs. Thus, if animals are pinealectomized prior to their exposure to the short days of winter, their gonads do not regress. Only after spring approaches do the gonads of intact animals recrudesce. Regrowth in the spring is a result of the onset of refractoriness to the antigonadotrophic influence of the pineal gland. The initiation of recrudescence of the sexual organs precedes the terminal arousal of the animals from hibernation. Following complete functional restoration of their reproductive systems, the animals emerge from hibernation, breed, and the females deliver their first litter soon thereafter. The refractoriness, which initiated gonadal recrudescence, extends into the summer period during which time it is eventually interrupted because of the exposure of the animals to alternating light:dark cycles for a several-month interval.

Thus, the photoperiod, because of its action on the pineal gland, is important in synchronizing the annual reproductive and hibernating cycles of the Syrian hamster. In the absence of the pineal gland, it may be argued that the animals would be continuous breeders and possibly never hibernate, both of which would be detrimental to survival of the species.

REFERENCES

1. Beckman, A. L. (1978) Hypothalamic and midbrain function during hibernation, in Current Studies on Hypothalamic Function 1978, Vol. 2, Veale, W. L. and Lederis, K., eds. Karger, Basel, pp. 29-43.

2. Benson, B. and Ebels, I. (1978) Pineal peptides. J. Neural Transmis., Suppl., 13, 157-173.

3. Czyba, J. C., Girod, C. and Durand, N. (1964) Sur l'antagonisme epiphyso-hypophysaire et les variations saisonnieres de la spermatogenese chez le Hamster dore (Mesocricetus auratus). C. R. Seanc. Soc. Biol., 158, 742-745.

4. Deane, H. W. and Lyman, C. P. (1954) Body temperature, thyroid and adrenal cortex of hamsters during cold exposure and hibernation, with comparisons to rats. Endocrinology, 55, 300-315.

5. Elliott, J. (1976) Circadian rhythms and photoperiodic time measurement in mammals. Fed. Proc., 35, 2339-2346.

6. Gaston, S., and Menaker, M. (1967) Photoperiodic control of hamster testis. Science, 158, 925-928.

7. Goodman, R. L. and Karsch, F. J. (1980) Control of seasonal breeding in the ewe: Importance of changes in response to sex-steroid feedback, in Seasonal Reproduction in Higher Vertebrates, Reiter, R. J. and Follett, B. K., eds. Karger, Basel, pp. 134-154.

8. Hall, V. and Goldman, B. (1980) Effects of gonadal steroid hormones on hibernation in the Turkish hamster (*Mesocricetus brandti*). J. Comp. Physiol., 135, 107-114.

9. Heller, H. C. (1979) Hibernation: Neural aspects. Ann. Rev. Physiol., 41, 305-321.

10. Heller, H. C. and Poulson, T. L. (1970) Circannian rhythms - II. Endogenous and exogenous factors controlling reproduction and hibernation in chipmonks (*Eutamias*) and ground squirrels (*Spermophilus*). Comp. Biochem. Physiol., 33, 357-383.

11. Hoffman, K. (1978) Photoperiodic mechanism in hamsters: The participation of the pineal, in Environmental Endocrinology, Assenmacher, I. and Farner, D. S., eds. Springer-Verlag, Berlin, pp. 94-102.

12. Hoffman, R. A. (1964) Terrestial animals in cold: Hibernators, in Handbook of Physiology Sec. 4, Dill, D. B., ed. American Physiological Society, Washington, D.C., pp. 379-403.

13. Hoffman, R. A. (1964) Speculations on the regulation of hibernation. Ann. Acad. Sci. Fenn., A, IV Biologica, 71, 201-216.

14. Hudson, J. W. (1973) Torpidity in mammals, in Comparative Physiology of Thermoregulation, Vol. III, Whittow, G. C., ed. Academic Press, New York, pp. 97-165.

15. Hudson, J. W. and Wang, L. C. H. (1979) Hibernation: Endocrinologic aspects. Ann. Rev. Physiol, 41, 287-303.

16. Johannson, B. W. (1978) Seasonal variations in the endocrine system of hibernators, in Environmental Endocrinology, Assenmacher, I. and Farner, D. S., eds. Springer-Verlag, Berlin, pp. 103-110.

17. Johansson, B. W. and Senturia, J. B. (1972) Seasonal variations in the physiology and biochemistry of the European hedgehog (*Erinaceus europaeus*) including comparisons with non-hibernators, guinea pig and man. Acta Physiol. Scand. Suppl., 380, 1-159.

18. Johnson, L. Y. and Reiter, R. J. (1978) The pineal gland and its effects on mammalian reproduction, in The Pineal and Reproduction, Reiter, R. J., ed. Karger, Basel, pp. 116-156.

19. Lyman, C. P. and Chatfield, P. O. (1955) Physiology of hibernation in mammals. Physiol. Rev., 35, 403-425.

20. Mogler, R. K. -H. (1958) Das endokrine System des Syrischen Goldhamsters (*Mesocricetus auratus auratus* Waterhouse) unter Berucksichtigung und experimentellen Winterschlafs. Zeit. Morph. Oekol. Tiere, 47, 267-308.

21. Reiter, R. J. (1969) Pineal-gonadal relationships in male rodents, in Progress in Endocrinology, Gual, C., ed. Excerpta Medica, Amsterdam, pp. 631-635.

22. Reiter, R. J. (1972) Evidence for refractoriness of the pituitary-gonadal to the pineal gland in golden hamsters and its possible implications in annual reproductive rhythms. Anat. Rec., 173, 365-371.

23. Reiter, R. J. (1973-1974) Influence of pinealectomy on the breeding capability of hamsters maintained under natural photoperiodic and temperature conditions. Neuroendocrinology, 13, 366-370.

24. Reiter, R. J. (1974) Pineal-mediated regression of the reproductive organs of female hamsters exposed to natural photoperiods during the winter months. Am. J. Obst. Gynec., 118, 878-880.

25. Reiter, R. J. (1974) Circannual reproductive rhythms in mammals related to photoperiod and pineal function: A review. Chronobiologia, 1, 365-395.

26. Reiter, R. J. (1975) The pineal gland and seasonal reproductive adjustments. Int. J. Biometeorol., 19, 282-288.

27. Reiter, R. J. (1975) Exogenous and endogenous control of the annual reproductive cycle in the male golden hamster: Participation of the pineal gland. J. Exp. Zool., 191, 111-119.

28. Reiter, R. J. (1980) The pineal and its hormones in the control of reproduction in mammals. Endocrinol. Rev., 1, 109-131.

29. Smit-Vis, J. H. (1972) The effect of pinealectomy and of testosterone administration on the occurrence of hibernation in adult male golden hamsters. Acta Morphol. Neerl. Scand., 10, 269-281.

30. Smit-Vis, J. H. and Smit, G. J. (1970) Hibernation and testes activity in golden hamster. Neth. J. Zool., 20, 502-506.

31. South, F. E., Breazille, J. E., Dellman, H. D. and Epperly, A. D. (1969) Sleep, hibernation and hypothermia in the yellow-bellied marmot (*M. flaviventris*), in Depressed Metabolism, Musacchia, X. J. and Saunders, A. J., eds. American Elsevier, New York, pp. 277-312.

32. Stetson, M. H., Matt, K. S. and Watson-Whitmyre, M. (1976) Photoperiodism and reproduction in golden hamsters: Circadian organization and the termination of photorefractoriness. Biol. Reprod., 14, 531-537.

33. Tucker, H. A. and Oxender, W. D. (1980) Seasonal aspects of reproduction, growth and hormones in cattle and horses, in Seasonal Reproduction in Higher Vertebrates, Reiter, R. J. and Follett, B. K., eds. Karger, Basel, pp. 155-180.

34. Van Horn, R. N. (1980) Seasonal reproductive patterns in primates, in Seasonal Reproduction in Higher Vertebrates, Reiter, R. J. and Follett, B. K., eds. Karger, Basel, pp. 181-221.

35. Vaughan, M. K. and Blask, D. E. (1978) Arginine vasotocin - A search for its function in mammals, in The Pineal and Reproduction, Reiter, R. J., ed. Karger, Basel, pp. 90-115.

36. Wetterberg, L. (1978) Melatonin in humans: Physiological and clinical studies. J. Neural Transmis. Suppl., 13, 289-310.

37. Wimsatt, W. A. (1969) Some interrelations of reproduction and hibernation in mammals, in Dormancy and Survival, Woolhouse, H. W., ed. Cambridge Univ. Press, Cambridge, pp. 511-559.

38. Zucker, I., Johnston, P. G. and Frost, D. (1980) Comparative, physiological and biochronometric analyses of rodent seasonal reproductive cycles, in Seasonal Reproduction in Higher Vertebrates, Reiter, R. J. and Follett, B. K., eds. Karger, Basel, pp. 102-133.

HUMORAL CONTROL OF HIBERNATION IN GOLDEN HAMSTERS

L. JANSKÝ, Z. KAHLEROVÁ, J. NEDOMA, AND J. F. ANDREWS[1]
Department of Comparative Physiology, Faculty of Science, Charles University, 128 44 Prague 2, Viničná 7, Czechoslovakia, and [1]Department of Physiology, Trinity College, Dublin, Ireland

INTRODUCTION

Although the golden hamster has been used as an experimental animal for more than 40 years, data concerning its hibernating capabilities are limited. Smit-Vis and Smit [27] observed that the length of the prehibernation period of golden hamsters was affected by season. They concluded that golden hamsters exposed to cold in November, December and January entered hibernation in February. At other seasons of the year they required at least a three-month period of cold exposure before they hibernated. On the other hand, Lyman and O'Brien [18], using a much greater number of animals, found no evidence that the tendency to hibernate was seasonal in this species.

There is a substantial amount of evidence showing that changes in the activity of hormonal glands and of concentration of hormones in the blood occur during the processes of hibernation [13]. However, direct observations on the effects of hormones on the hibernating capabilities of the golden hamster have not been made. Only Smit-Vis [26] and Smit-Vis and Smit [28] have studied the role of the pineal and the gonads in the hibernation of the golden hamster.

The aim of this report is to contribute to understanding the role of some hormonal glands (namely the pineal, gonads, adrenals and thyroid) in the control of hibernation in the golden hamster. Seasonal features of hibernation and the effect of two external factors, light and temperature, on hibernation also are presented. Much of this work was stimulated by the hypothesis [14] that changes in function of the gonads and thyroid might be related to the greatly increased metabolism of serotonin (5-HT) known to occur in the brain of hibernating animals [21].

EXPERIMENTAL PROCEDURES

Animals: Young male golden hamsters, average body weight about 100 gms, were divided into nine groups of 15-30 individuals each. They served as controls for other groups of animals in which the hormonal status was changed experimentally. Before performing any tests, the animals were kept for at least

14 days in animal rooms at a constant temperature of 20°C on a 12:12 hours light-dark cycle. The individual groups of animals were then placed in either continuous darkness or continuous light in a cold room located in a deep basement. Under these conditions acoustic and other disturbances were avoided. The average temperature in this hibernation room varied from 5.3° to 15.3°C, depending on the season of the year. The temperature of the hibernation room did not show any daily fluctuation, but rather changed gradually during the course of the experiment (Fig. 1). All animal groups, unless otherwise stated, were kept undisturbed in the dark for 60 days and were fed on grain and carrots. The periods of hibernation were determined by daily examinations at 11a.m. using a red flash light.

Methods: In experimental groups of golden hamsters, pinealectomy, thyroidectomy or castration were performed by standard procedures under pentobarbital (Spofa) anesthesia. The animals were allowed to recover for about a week before being exposed to cold. Animals in control groups were sham-operated. The effectiveness of the extirpation was examined postmortem.

Various drugs in crystaline form contained in silastic tubing (o.d. 1.96 mm, i.d. 1.70 mm, Dow-Corning Corp., Midland, Michigan, U.S.A) were implanted under the skin in some animals and in others triiodothyronine (rT_3) was administered in saline. Prior to implantation the tubing was filled with the drug and was sealed at both ends by Silastic 382 Medical Grade Elastomer (Dow Corning Corp., Midland, Michigan, U.S.A.). This procedure allowed continuous and long-term administration of drugs to undisturbed animals as the drugs slowly leached through the silastic walls. Preliminary analyses indicated that the daily release of melatonin was about 5 µg/animal/1 cm of tubing. Empty tubing was implanted under the skin in control groups. An outline of treatments is presented in Table 1.

In all groups of animals the weight of both testes and adrenals was estimated at the termination of experiments in those individuals showing the most frequent hibernation pattern.

Criteria of hibernation: Before studying the role of hormonal mechanisms in the control of hibernation, the hibernating capabilities of normal golden hamsters were tested at different seasons of the year. These studies, however, were complicated by the fact that as yet exact criteria have not been established for hibernating capability. Lyman and O'Brien [18] suggested that at least five interdependent *factors* should be considered: "1. the number of animals in a group which hibernate during the period of exposure to cold, 2. the span of time between cold exposure and hibernation, 3. the length of the hiber-

HIBERNATION OF CONTROL GOLDEN HAMSTER

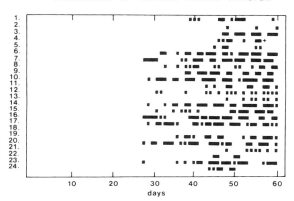

Experiment was started on October 2, 1978

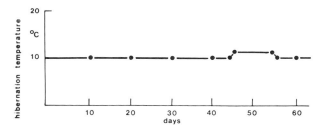

Fig. 1. Hibernation pattern of golden hamsters exposed to cold and constant dark (bars = hibernating). Temperature in the hibernation room is given in the lower figure.

hibernation during the 60-day period (similar to *factor* 2). Only data from animals surviving the entire experiment were used.

As supporting criteria the following parameters were used: (1) The percentage of animals hibernating during the two-month stay in the cold (similar to *factor* 1). (2) The hibernation bout (similar to *factor* 5). (3) The time when the first animal in a group started to hibernate (H_1).

TABLE 1

DRUG AND TREATMENT PROCEDURES IN GOLDEN HAMSTERS

substance	source	amount mg	length of tubing mm
melatonin	Sigma Chem. Corp., St. Louis, U.S.A.	50	50
testosterone propionate	Spofa, Prague, C.S.S.R.	3 or 13	10
reverse triiodothyronine (rT$_3$)	Henning, Berlin, West Germany	0.065	10
combination:			
fluoxetine HCl*	Lilly Res. Lab., Indianapolis, U.S.A.	3.2	
pargyline	Abbott Labs., North Chicago, U.S.A.	5.2	20
α-methyl-p-tyrosine	Sigma Chem. Corp., St. Louis, U.S.A.	6.5	
desoxycortico-sterone	Spofa, Prague C.S.S.R.	3.6	5
cortisol	Spofa, Prague, C.S.S.R.	3.3	5

*The authors are indebted to Lilly Res. Lab., Indianapolis, U.S.A., for the free supply of fluoxetine.

nating season, 4. the amount of time spent in hibernation compared to the amount of time in the cold, and 5. the length of the period during which an animal remains in continuous hibernation."

The relative importance of these individual criteria for defining hibernation was not evaluated by these authors, but Lyman and O'Brien [18] indicated "the maximum time of uninterrupted hibernation is more definitive than an average of the lengths of the bouts of hibernation."

In our experiments the following criteria were found to be the most informative for defining hibernation: (1) Frequency of hibernation expressed as total time spent in hibernation by all animals in the group during the two-month stay in the cold divided by the number of animals. This is similar to the Lyman and O'Brien *factor* 4. (2) The prehibernation period (PHP) considering only hibernating animals of the group. This was expressed as the accumulated total of days prior to hibernation divided by the number of animals which entered

The length of the hibernation season, Lyman and O'Brien's *factor* 3, was not studied in detail in our experiments. However, we also consider that this factor would be significant in clarifying the hibernating capabilities of animals.

RESULTS

Seasonality of hibernation and the possible effect of environmental temperature: A typical course of hibernation in one group of golden hamsters is shown in Fig. 1. It is evident that a relatively long period of cold exposure must occur before the animals are able to enter hibernation. (All season average = 43.6 days.) Once hibernation starts it occurs in typical bouts of relatively short periods of duration, one to seven days.

Eight experimental groups of animals were used to study the seasonality of hibernation. The mortality during the experiments was about 5%, and the loss in body weight was 20%. Golden hamsters were able to hibernate at any time of the year. The percentage of nonhibernating animals during a two-month cold exposure varied considerably in different seasons, being relatively high in the late spring (Fig. 2). Of 171 golden hamsters used, almost 72% entered hibernation during the two-month stay in the cold (Table 2). Additional experiments have shown that during a longer stay in the hibernation room all animals finally hibernated. Lyman and O'Brien [18] observed that 68% of golden hamsters hibernated at 5°C. Thus, in our view the golden hamster is considered to be a good hibernator rather than a poor one.

The average length of the prehibernation period did not show a distinct seasonal pattern, but seemed to be the longest in late spring (49.8 days) and the shortest in late February (34.6 days) (Fig. 2). Depending on the group, individual animals (H_1) started to hibernate for the first time after a period of 12-43 days of cold exposure. A positive correlation was demonstrated between average environmental temperature in the hibernation room and length of the prehibernation period (PHP) (Fig. 3).

Similarly, the frequency of hibernation, although it seemed to be lowest in late spring and highest in the fall, did not show a well-defined seasonal pattern (Fig. 2). However, there was a suggestion that a negative correlation existed between environmental temperature and frequency of hibernation. Time spent in hibernation varied from 14% in October to 0.9% in May. These results are close to the range reported by Lyman and O'Brien [18] (11.3-13.7%).

In general, we can say that PHP and hibernation bouts are less affected by

TABLE 2
HIBERNATION OF CONTROL GROUPS OF GOLDEN HAMSTERS AT DIFFERENT MONTHS OF THE YEAR

groups	n	% deaths	% non-hib.	PHPa (days)	H1b	freq. of hib.c	hib. bout (days)	testes (mg)d	adrenals (mg)d	hib. room temp (°C)e
controls-dark 16.2.79f	35	2.9	20.0	37.2±14.7	6	5.8	1.6	n=6 1060±600	n=6 27.0±7.3	9.6
controls-dark 26.2.80	14	7.1	30.8	34.6±14.1	16	6.9	1.6	n=7 925±270	n=7 21.6±3.0	9.3
controls-dark 15.5.79	15	0.0	73.3	49.8±7.8	38	0.5	1.0	n=6 510±143	n=6 19.7±3.2	15.3
controls-dark 1.9.78	15	0.0	13.3	41.6±5.8	32	7.7	1.5	n=9 674±240	n=9 22.4±2.9	10.0
controls-dark 2.9.76 (cornfed)	30	3.3	20.7	47.7±4.5	43	4.9	---	---	---	12.1
controls-dark 2.10.78	25	8.0	4.3	36.7±7.9	28	14.0	1.9	n=6 515±130	n=6 21.4±4.0	10.2
controls-dark 16.10.79	17	5.9	25.0	43.6±10.8	26	4.6	1.6	n=8 294±90	n=8 22.8±5.7	11.1
controls-dark 22.12.79	20	0.0	16.7	36.7	11	3.6	1.4	---	---	7.8
controls-dark 22.12.79♀	18	11.1	18.8	31.7	11	5.4	1.5	---	---	7.8
controls-light 26.2.80	12	8.3	54.5	48.6±10.1	32	2.5	1.3	n=5 836±320	n=5 26.7±2.0	10.2
% of controls in darkg			+76.95	+40.46		-63.77		-9.63	+23.61	
controls-light 16.10.79	20	5.0	60.0	34.7±11.1	22	1.9	1.4	n=8 579±170	n=8 27.1±3.5	12.7
% of controls in dark			+140.00	-20.05		-59.58		+96.94	+18.86	

aPHP = prehibernating period; bH1 = day when first animal in group started hibernating; c = days/animal; dexpressed per 100 gm of body weight; etwo-month average; fDate of the beginning of experiment; gcontrols = 100%

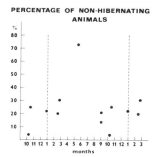

PERCENTAGE OF NON-HIBERNATING ANIMALS

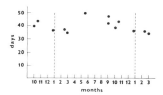

PREHIBERNATION PERIOD

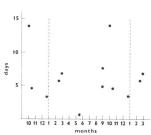

FREQUENCY OF HIBERNATION

Fig. 2. Percentage of nonhibernating individuals, duration of the prehibernation period and frequency of hibernation of groups of golden hamsters studied at different times of the year. Months are designated by number, e.g., January = 1, February = 2, etc.

different seasons of the year and are more temperature dependent than the frequency of hibernation.

Our results indicate that in golden hamsters there are no prerequisite qualitative changes for the occurrence of hibernation at different seasons of the year. The animals are genetically predisposed for hibernation at all seasons. The observed variability in hibernation capabilities throughout the

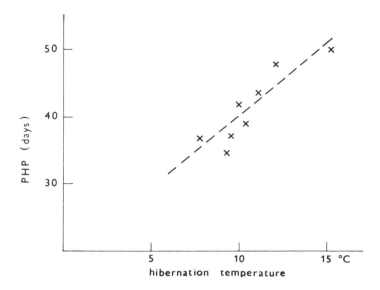

Fig. 3. Effect of hibernation temperature on the duration of the prehibernation period (PHP) of individual groups of male golden hamsters studied at different seasons of the year.

year can be explained either by the effect of environmental temperature or by the different physiological status of the animals during the experiments, e.g., the great differences in the weights of the testes (265-933 mg in different groups). Therefore, in future studies the physiological status of animals should be carefully standardized and animals should be held for longer than two weeks under constant conditions prior to any experiments.

Sexual differences: The hibernation capabilities of one group of female golden hamsters (N = 18) were compared with that of males (N = 20). Both groups were placed in the hibernation room in December 1979. Data presented in Table 2 indicate that females hibernated for 50% longer than males. The PHP was not changed significantly in females.

Effect of light on hibernation: Two experiments were performed, one in February and the other in October. Under continuous light the percentage of nonhibernating animals increased by 77% in February and by 140% in October as compared with continuously dark exposed animals at the same season. The PHP tended to be longer in February and shorter in October. These changes were not significant, however. Frequency of hibernation fell to 36% and 40% in February

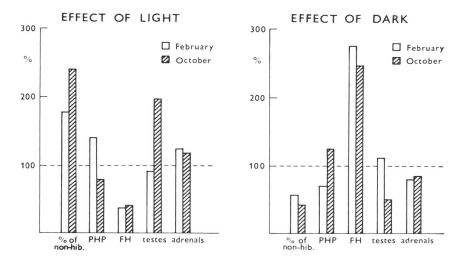

Fig. 4. Hibernating capabilities and changes in the weights of testes and adrenals of golden hamsters exposed to continuous light or darkness. Data are expressed as percentage of nonhibernating, prehibernating period (PHP) and frequency of hibernation (FH) of animals kept in continuous light or continuous dark.

and October, respectively, and the duration of a hibernation bout was somewhat shortened compared with totally dark exposed animals. A great increase in the weight of testes was observed in October (97%), while in February no significant change occurred. Light exposed animals showed a marked increase in adrenal weight in both months (23.6% and 18.9%, respectively) (Fig. 4 and Table 2).

Conversely, when expressing the data in terms of light exposed animals, we can say that animals kept in continuous darkness hibernate considerably better than those kept in continuous light. The percentage of nonhibernating individuals decreased to 56.5% in February and to 41.7% in October. Frequency of hibernation increased by 176% in February and by 147% in October. The improved hibernating capabilities in dark exposed animals are followed by a decrease in the weight of both adrenals and testes (Fig. 4).

Hormonal mechanisms mediating the effect of darkness on hibernation: Further experiments were designed to find out whether hormonal mechanisms are responsible for the positive effect of darkness on hibernation. It has been well demonstrated that in small mammals the effect of darkness is mediated by the pineal, which inhibits the activity of the gonads by means of an increased production of melatonin [11, 12, 16, 24].

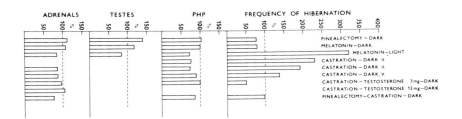

Fig. 5. Effect of pinealectomy, castration, melatonin and testosterone on hibernation of golden hamsters and on weights of testes and adrenals.

Therefore, the effect of pinealectomy was studied on cold exposed golden hamsters kept in the dark. The frequency of hibernation was lowered by 20% after pinealectomy, compared with controls kept under similar conditions at the same season. The weight of the testes increased 52% and the weight of the adrenals also appeared to be positively affected (Fig. 5, Table 3).

The results show that pinealectomized animals retained the capacity to hibernate, although to a somewhat lesser extent. It can therefore be assumed that the activity of the pineal can influence hibernation, but it is not the only mechanism involved. A marked increase in the weight of the testes showed the effectiveness of this treatment and confirmed the functional relationship between the pineal and gonads.

In order to clarify in more detail the mechanisms mediating the effect of dark on hibernation, further experiments were performed with melatonin implanted under the skin followed by subsequent exposure of golden hamsters to cold and darkness. In agreement with Reiter et al. [24], melatonin administered to animals kept in the dark (when pineal activity was probably already increased) induced a paradoxical effect. It resulted in slight increases in weight of the testes (29.1%) and adrenals (3.3%) and lowered the frequency of hibernation by 20% (Fig. 5, Table 3). The duration of PHP was unchanged.

On the other hand, in animals kept in constant light, when low pineal activity can be expected, melatonin lowered the weight of testes and adrenals (16.2% and 15.1%, respectively) and increased the frequency of hibernation by 221%. Thus, we can conclude that melatonin was very effective in inducing hibernation and suppressing gonadal activity. Lynch [19] also observed that after melatonin treatment white-footed mice became torpid 2.5 times more frequently with a long-day photoperiod than with a short-day photoperiod.   Palmer and Riedesel [23] were able to increase the incidence of hibernation in ground squirrels after daily injections of melatonin.

These findings support the primary hypothesis that the increased hibernating capabilities of golden hamsters, when exposed to cold and to constant dark, might be due to an increased production of melatonin from the pineal and/or due to the subsequent involution of the gonads, accompanied by the tendency for involution of adrenals.

These data, however, did not clarify the relative importance of the pineal and of the testes in induction of hibernation.   Therefore, experiments were designed to study the significance of gonadal activity   in hibernation. Experiments were performed on three groups of castrated animals exposed to cold and to continuous darkness. Fig. 5. demonstrates that castrated animals kept in the cold and dark hibernated much better than controls.   PHP was shortened and the frequency of hibernation increased by 131% and by 92.7% in February and by 37.7% in May. The decreased weight of the adrenals was not significant.

Thus, it can be concluded that lowered gonadal activity occurrs in parallel with increased frequency of hibernation.   This was confirmed by other experiments in which the gonadal activity was simulated by administration of testosterone.   Doses of testosterone, from silastic tubes implanted under the skin, were applied to castrated animals.   The animals were then exposed to cold and darkness.   As is evident in Fig. 5, a small dose of testosterone (3 mg) lowered the frequency of hibernation by 49.1%, compared to controls, and by 67% compared to castrated individuals.   A higher dose, 12 mg, prevented hibernation completely.   The duration of the PHP was not changed after the small dose of testosterone.   In both cases the weight of adrenals was not changed, although in other experiments the lowered frequency of hibernation was usually accompanied by an increased weight of the adrenals.

The inhibitory effect of testosterone and increased gonadal activity on hibernation has already been observed [10, 17, 28].

These results confirmed the importance of gonadal activity in the control of hibernation.   The question still remained, however, whether or not the gonadal activity was the only factor involved. Therefore, further experiments were performed on castrated and pinealectomized animals exposed continuously to cold and darkness.   Results showed that this treatment had a negative effect on hibernation (Table 3, Fig. 5).   The frequency of hibernation, which was shown to increase in castrates, remained at the level observed in controls and the duration of PHP was not significantly shortened.   There was a decrease in the

TABLE 3

EFFECT OF PINEALECTOMY, CASTRATION AND ADMINISTRATION OF MELATONIN
AND TESTOSTERONE ON HIBERNATION OF GOLDEN HAMSTERS

group	n	% deaths	% non-hib.	PHPa (days)	H1b	freq. of hib.c	hib. bout (days)	testes (mg)d	adrenals (mg)d	hib. room temp. (°C)e
pinealectomy								n=6	n=6	
dark 2.10.78f	25	8.3	9.1	37.3+9.0	24	11.4	1.5	783+15	24.0+10.9	10.4
% of controlsg				+1.63		-18.58		+52.0		
melatonin-dark								n=7	n=7	
2.10.78	24	8.3	4.2	37.9+61	28	11.2	1.6	665+18	22.1+0.3	10.4
% of controls				+3.27		-20.00		+29.1	+3.3	
melatonin-light								n=8	n=8	
16.10.79	19	10.5	42.1	25.8+15.1	14	6.1	1.4	485+16	23.0+0.9	12.7
% of controls				-25.70		+221.05		-16.2	-15.1	
castration-dark										
15.2.79	14	29.0	10.0	35.8+10.1	25	13.4	1.6	---	---	9.7
% of controls				-23.50		+131.03				
castration-dark									n=7	
26.2.80	15	6.7	14.3	25.9+14.3	7	13.3	1.6	---	18.7+0.3	
% of controls				-25.15		+92.75			-13.40	

aPHP = prehibernating period; bH1 = day when first animal in group started hibernating; c = days/animal; dexpressed per 100 gm of body weight; etwo-month average; fdate of the beginning of experiment; gcontrols = 100%

TABLE 3 Continued

group	n	% deaths	% non-hib.	PHPa (days)	H1b	freq. of hib.c	hib. bout (days)	testes (mg)d	adrenals (mg)d	hib. room temp. (°C)e
castration-dark										
15.5.79f	15	0.0	80.0	45.0+6.9	41	0.73	1.0	---	n=6 17.2+2.7	15.3
% of controlsg				-9.00		+37.74			-12.69	
castration-dark testosterone 3 mg										
15.5.79	15	0.0	86.7	50.0+4.2	47	0.27	1.0	---	n=6 19.1+2.2	15.3
% of controls				+0.40		-49.06			-3.05	
castration-dark testosterone 12 mg										
15.5.79	15	7.0	100.0	∞		∞	0.0	---	n=6 20.9+2.9	15.3
% of controls									+6.09	
pinealectomy-castration-dark										
26.2.80	13	7.1	41.7	30.1+16.1	14	6.8	1.5	---	n=4 16.7+1.6	9.3
% of controls				-13.00		-1.45			-22.7	

aPHP = prehibernating period; bH$_1$ = day when first animal in group started hibernating; c = days/animal; dexpressed per 100 gm of body weight; etwo-month average; fdate of the beginning of experiment; gcontrols = 100%

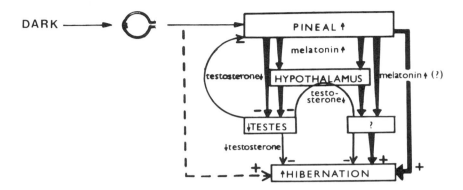

Fig. 6. A schematic interpretation of mechanisms controlling hibernation in continuous darkness.

weight of adrenals as compared with controls (22.1%). Thus, it can be concluded that pinealectomy prevents the positive effect of castration on hibernation.

There are many interrelationships and a summary view of these data is presented in Fig. 6.

In animals exposed to dark the pineal gland is activated and exerts a very strong positive influence on hibernation. This is related to the lowered gonadal activity. Since the pineal also affects hibernation in castrated animals, its influence is probably brought about by other routes than the gonads. The effect of the pineal on hibernation can be either direct or indirect, being mediated by the hypothalamus or by some other endocrine gland. It is not yet known which humoral factor released by the pineal affects hibernation positively. It might be melatonin, but other substances cannot be excluded.

Testosterone, on the other hand, inhibits hibernation. Whether or not this inhibitory effect is direct or is brought about through the pineal, the hypothalamus or by other routes remains to be seen. In addition, since hibernation takes place in the absence of both the pineal and the gonads, some extrapineal and extragonadal mechanisms must exist which contribute to the occurrence of hibernation. Data also indicate that the frequency of hibernation, rather than the length of PHP reflects changes in the hormonal status of animals.

The role of adrenals in hibernation: In addition to the pineal, changed activity of the adrenals may also influence the hibernation behavior of golden hamsters to a certain extent. This is evident from our data which indicated that adrenal weights usually tended to decrease when gonadal weight was lowered

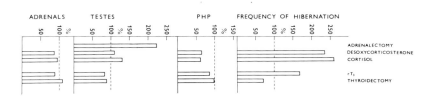

Fig. 7. Effect of adrenalectomy, thyroidectomy and some drugs on hibernation of golden hamsters and on weights of testes and adrenals.

and, at the same time, the hibernating capabilities increased. Lowered adrenal activity during hibernation is also indicated in published results which showed no change in adrenal activity or concentration of adrenal steroids in the blood during hibernation [7, 8, 25] or during exposure to cold [14].

Contrary to these published data, our experiments have shown that the adrenals are very important for the induction of hibernation. It was found that adrenalectomy prevents hibernation completely and at the same time causes an increase in testicular weight (Table 4, Fig. 7). Whether the effect of the absence of adrenal hormones acts directly in preventing hibernation, or is brought about by causing the increased activity of the gonads remains to be seen.

It was shown that subcutaneous administration of two steroid substances (a mineralocorticoid, desoxycorticosterone, and a glucocorticoid, cortisol) to intact animals increased the frequency of hibernation by 146% and 160%, respectively (Table 4, Fig. 7).

Furthermore, in recent experiments, Musacchia and Deavers (this volume) maintain that the glucocorticoid, cortisone, is involved in carbohydrate maintenance and has a role in thermogenesis in the hypothermic golden hamster. Their data indicate that the adrenal cortex has a significant role in hibernation.

The role of the thyroid in hibernation: Another hormonal system must be considered as possibly involved in the control of hibernation, namely, the thyroid gland. As with the other systems, the activity of this gland during hibernation has not yet been clarified [13]. Interaction of thyroid with other hormonal systems influencing hibernation has been demonstrated. For example, plasma thyroxine levels are lowered after treatment with melatonin [30]. In our experiments thyroidectomy, performed two to three weeks prior to placing the

TABLE 4

EFFECT OF ADRENALECTOMY, THYROIDECTOMY AND SOME DRUGS ON HIBERNATION OF GOLDEN HAMSTERS

group	n	% deaths	% of nonhib. animals	PHPa (days)	H1b	freq. of hib (days/anim.)	hib. bout (days)	testes (mg)c	adrenals (mg)c	hibernation room temperature (°C)d
adrenalectomy-dark 16.2.79e	24	75.0	100.0	∞	∞	---	---	n=6 2070+1.21	---	9.6
% of controlsf								+123		
desoxycortico-sterone-dark 16.10.79	19	0.0	5.3	28.6+10.2	14	11.1	1.6	n=8 324+103	n=8 19.8+3.9	11.1
% of controls				-34.41		+136.17		+10-20	-13.10	
cortisol-dark 16.10.79	17	5.9	6.3	26.8+11.1	8	12.2	1.7	n=8 388+101	n=8 21.7+2.0	11.1
% of controls				-38.54		+159.57		+31-97	-4.83	
rT3-dark 15.5.79	15	88.0	0.0	42.7+7.4	44	0.84	1.4	427+163	17.4+3.2	15.3
% of controls				-14.46		+68.0		-16.3	-11.68	
thyroidectomy-dark 1.9.78	20	0.0	20.0	41.5+7.7	31	5.5	1.6	n=9 604+277	n=9 24.5+4.4	10.0
% of controls				-0.72		-28.57		-10-38	+9.375	
combination (pargyline + fluoxetine + α-MPT)-dark	17	15.0	11.8	36.6	25	10.5	1.6	597+67	24.5+3.1	11.1
% of controls				-15.67		+123.40		+108.01	+9.86	

aPHP = prehibernating period; bH1 = day when first animal in group started hibernating; cexpressed per 100 gm of body weight; dtwo-month average; edate of beginning of experiment; fcontrols = 100%

animals in cold and darkness, lowered the tendency for hibernation. Thus, a certain level of thyroid activity is probably necessary for successful hibernation.

It has recently been shown that a thyroid hormone, reverse triiodothyronine (rT_3) occurs in small quantities in the blood of normal animals. The relative concentration of this hormone increases under certain conditions [2]. It has been suggested that this hormone may act as a competitive inhibitor of the powerful metabolic stimulant T_3 at specific sites within the cell. We have tested the effect of this hormone on hibernation by subcutaneous implantation of silastic tubing containing rT_3 in saline. The prehibernating period was unaffected but there was a 68% increase in the number of hibernating animals. Together with the effect of thyroidectomy, which would reduce availability of both T_3 and rT_3, these findings would also point to a role for the thyroid in the processes of hibernation.

The role of serotonin in the control of hibernation: Previously, we have shown that in hibernating golden hamsters the turnover of serotonin (5-HT) in the brain stem is greatly increased [21]. The hypothesis was elaborated that increased activity of serotonergic pathways in the brain plays a role in controlling the hypothalamic hormones regulating the activity of other hormonal glands [14, 15].

The important role of 5-HT metabolism in the control of hibernation was further indicated in our experiments which showed that an increased supply of tryptophane, the precursor of 5-HT, causes an increase in the occurrence of hibernation in golden hamsters [16]. More recently we found that administration of a combination of drugs which stimulates 5-HT metabolism (fluoxetine HCl, parqyline) and simultaneously inhibits norepinephrine metabolism (α-methyl-p-tyrosine-- α-MPT) stimulates hibernation (Table 4). The frequency of hibernation was increased by 123%.

Thus, the role of 5-HT metabolism in the control of hibernation must also be considered very important. Serotonergic pathways may exert their influence on the hypothalamus or on the pineal or may affect hibernation by other routes.

Hibernation trigger in the golden hamster: Finally, a possible role for hibernation trigger cannot be excluded from our consideration of factors influencing hibernation. This substance, identified as a proalbumin [22] actively promotes hibernation in seasonal hibernators, marmots and ground squirrels [3, 4, 5, 6, 29]. On the other hand, hibernation trigger has not been found to be active in Richardson's ground squirrels [1, 9]. In golden hamsters the dialysate from the blood of hibernating animals does not effect the prehibernation period [20].

Some preliminary experiments performed by Dr. Marek at the University of Brno (unpublished) indicate the presence of a proalbumin factor in the blood of hibernating golden hamsters (cellulose acetate paper electrophoresis, 300 V, 2 mA, 15 min Veronal buffer, pH 8.6). Thus, the existence of a hibernation trigger in the golden hamster cannot be excluded.

SUMMARY

Golden hamsters are able to hibernate at any season of the year. Successful entrance into hibernation is much enhanced by total absence of light and by exposure to low temperature. Under these conditions the average prehibernation period lasts about six weeks. The frequency of hibernation, expressed as days one animal hibernates during a two-month period, is a better indicator of hibernation capabilities under different experimental conditions than is the span of the prehibernation period.

The experimental data from our laboratory indicate that hibernation in golden hamsters is under multiple hormonal control. In the dark the pineal exerts a very strong positive influence on hibernation. This is associated with lowered gonadal activity, but the effect of the pineal is also brought about by other routes, probably due to direct or indirect effects of melatonin. In addition, other hormonal mechanisms also play a role. Hibernation requires the presence of both the adrenals and the thyroid. Adrenal steroids and rT_3 have a positive effect on hibernation. Hibernation is also under the control of serotonergic pathways in the brain.

REFERENCES

1. Abbotts, B., Wang, L. C. H. and Glass, J. D. (1979) Absence of evidence for a hibernation "trigger" in blood dialysate of Richardson's ground squirrel. Cryobiology, 16, 179-183.

2. Andrews, J. F., Vybiral, S., O'Connor, M., Dennehy, A. and Cullen, M. (1978) Time course of minimum metabolic rate increase correlated with changes in thyroxine, triiodothyroxine and reverse triiodothyronine levels in the neonatal lamb, in New Trends in Thermal Physiology, Houdas, Y., and Guieu, J. D., eds. Mason, Paris, 58-61.

3. Dawe, A. R. and Spurrier, W. A. (1969) Hibernation induced in ground squirrels by blood transfusion. Science, 163, 298-299.

4. Dawe, A. R. and Spurrier, W. A. (1972) The bloodborne trigger for natural mammalian hibernation in the 13-lined ground squirrel and the woodchuck. Cryobiology, 9, 163-172.

5. Dawe, A. R. and Spurrier, W. A. (1974) Summer hibernation of infant (six-week old) 13-lined ground squirrels, *Citellus tridecemlineatus*. Cryobiology, 11, 33-43.

6. Dawe, A. R., Spurrier, W. A. and Armour, J. A. (1970) Summer hibernation induced by cryogenically preserved blood "trigger." Science, 168, 497-498.

7. Deane, H. W. and Lyman, C. P. (1954) Body temperature, thyroid and adrenal cortex of hamsters during cold exposure and hibernation, with comparisons to rats. Endocrinology, 55, 300-315.

8. Denyes, A. and Horwood, R. H. (1960) A comparison of free adrenal cortical steroids in the blood of a hibernating and non-hibernating mammal. Can. J. Biochem. Physiol., 38, 1479-1487.

9. Galster, W. A. (1978) Failure to initiate hibernation with blood from the hibernating Arctic ground squirrel, *Citellus undulatus*, and Eastern woodchuck, *Marmota monax*. J. Therm. Biol., 3, 93.

10. Hall, V. and Goldman, B. (1980) Effects of gonadal steroid hormones on hibernation in the Turkish hamster (*Mesocricetus brandti*). J. Comp. Physiol., 135, 107-114.

11. Hoffman, R. A., Hester, R. J. and Towns, C. (1965) Effects of light and temperature on the endocrine system of the golden hamster (*Mesocricetus auratus* Waterhouse). Comp. Biochem. Physiol., 15, 525-533.

12. Hoffmann, K. (1979) Photoperiod, pineal, melatonin and reproduction in hamsters, in The Pineal Gland of Vertebrates including Man, Ariëns Kappers, J., and Pevet, P., eds. Elsevier/North-Holland Biomedical Press, Amsterdam, Vol. 52, 397-415.

13. Hudson, J. W. and Wang, L. C. H. (1979) Hibernation: Endocrinologic aspects. Ann. Rev. Physiol., 41, 287-303.

14. Janský, L. (1978) Time sequence of physiological changes during hibernation: The significance of serotonergic pathways, in Strategies in Cold: Natural Torpidity and Thermogenesis, Wang, L. C. H. and Hudson, J. W., eds. Academic Press, New York, 299-325.

15. Janský, L. and Novotna, R. (1976) The role of central aminergic transmission in thermoregulation and hibernation, in Regulation of Depressed Metabolism and Thermogenesis, Jansky, L. and Musacchia, X. J., eds. Charles C Thomas, Springfield, Illinois, 64-80.

16. Johnson, L. Y. and Reiter, R. J. (1978) The pineal gland and its effects on mammalian reproduction. Prog. Reprod. Biol., 4, 116-156.

17. Lyman, C. P. and Dempsey, E. W. (1951) The effect of testosterone on the seminal vesicles of castrated, hibernating hamsters. Endocrinology, 49, 647-651.

18. Lyman, C. P. and O'Brien, R. C. (1977) A laboratory study of the Turkish hamster *Mesocricetus brandti*. Breviora Mus. Comp. Zool., 442, 1-27.

19. Lynch, G. R. (1977) Photoperiod, melatonin and spontaneous daily torpor, in Strategies in Cold: Natural Torpidity and Thermogenesis, Wang, L. C. H. and Hudson, J. W., eds. Academic Press, New York, 299-325.

20. Minor, J. G., Bishop, D. A. and Badger, C. R., Jr. (1978) The golden hamster and the blood-borne hibernation trigger. Cryobiology, 15, 557-562.

21. Novotna, R., Janský, L. and Drahota, Z. (1975) Effect of hibernation on serotonin metabolism in the brain stem of the golden hamster (*Mesocricetus auratus*). Gen. Pharmac., 6, 23-26.

22. Oeltgen, P. R., Bergmann, L. C., Spurrier, W. A. and Jones, S. B. (1978) Isolation of a hibernation inducing trigger(s) from the plasma of hibernating woodchucks. Prepar. Biochem., 8, 171-188.

23. Palmer, D. L. and Riedesel, M. L. (1976) Responses of whole-animal and isolated hearts of ground squirrels, *Citellus lateralis*, to melatonin. Comp. Biochem. Physiol., 53C, 69-72.

24. Reiter, R. J., Rollag, M. D., Panke, E. S. and Banks, A. F. (1978) Melatonin: Reproductive effect. J. Neural Transmiss., Suppl. 13, 209-223.

25. Schindler, W. J. and Knigge, K. M. (1959) Adrenal cortical secretion by the golden hamster. Endocrinology, 65, 739-747.

26. Smit-Vis, J. H. (1972) The effect of pinealectomy and of testosterone administration on the occurence of hibernation in adult male golden hamsters. Acta Morph. Neerl.-Scand., 10, 196-282.

27. Smit-Vis, J. H. and Smit, G. J. (1963) Occurrence of hibernation in the golden hamster, *Mesocricetus auratus* Waterhouse. Experientia, 19, 363-364.

28. Smit-Vis, J. H. and Smit, G. J. (1970) Hibernation and testes activity in the golden hamster. Netherl. J. Zool., 20, 502-506.

29. Spurrier, W. A., Folk, G. E., Jr. and Dawe, A. R. (1976) Induction of summer hibernation in the 13-lined ground squirrel shown by comparative serum transfusions from Arctic mammals. Cryobiology, 13, 368-374.

30. Vriend, J. and Reiter, R. J. (1977) Free thyroxin index in normal, melatonin-treated and blind hamsters. Horm. Metab. Res., 9, 231-234.

ROLE OF THE ENDOCRINE GLANDS IN HIBERNATION WITH SPECIAL REFERENCE TO THE
THYROID GLAND

JACK W. HUDSON
Department of Biology, University of Alabama in Birmingham, University Station,
Birmingham, Alabama 35294, U.S.A.

INTRODUCTION

One of the most bewildering aspects of the physiology of hibernation in
mammals has been trying to understand the role of the endocrine glands.
Although there have been a great many studies dealing with the status of the
endocrine glands in hibernation, most of which have been recently reviewed [16],
unequivocal answers to ill-defined questions remain elusive. Depending upon the
species of mammal, the gland selected for study, and the technique used to
assess the functional status of a particular gland, it is possible to promulgate
and defend contradictory hypotheses, i.e., the endocrine glands must become
inactive prior to hibernation or the endocrine glands are functional at the time
of hibernation so that any changes correlated with hibernation are an effect of
low body temperature.

It has been a long-standing assumption that endocrine inactivity is a necessary preparation for hibernation [9]. This premise is reinforced by many
observations of endocrine inactivity during hibernation [20], though most of
these early studies depended on light-microscopic histology of glands to infer
levels of activity.   However, hypothermia can elicit some of the same endocrine
hypofunctions observed histologically in the endocrine glands of hibernators in
hibernation [5].   Endocrine inactivity during hibernation may, in fact, be the
consequence of low body temperature.

In reviewing the literature, it is often difficult to discern the precise
nature of questions being asked.   Must there be involution of the endocrine
glands before an animal can hibernate?   If so, what is the purpose of involution?   Do the endocrine glands function during hibernation, but at a reduced
level, or are they completely inactive throughout the hibernation season even
though animals periodically arouse? Does the neuroendocrine system regulate the
switching on or switching off of the endocrine secretory functions?   Do the
endocrine glands, which have been inactive throughout hibernation, become active

prior to the termination of hibernation? Are some glands active during hibernation while others are inactive?

Much of the hibernation literature describes the physiology of bats, hamsters, hedgehogs, marmots, dormice and ground squirrels. The literature describes the typical hibernator as one which lowers its body temperature periodically to depths approaching freezing throughout the winter season. However, from the study of a wide variety of mammals has come the realization that hibernation is merely one end of a continuum of thermoregulatory performances, with shallow daily torpor at the other end [12]. By analogy, it seems reasonable that endocrine function may differ among various kinds of hibernators: complete endocrine inactivity observed in deep hibernators and endocrine function continuing at a reduced level in shallow hibernators. Historically there has been the tendency to assume that observations made on ground squirrels, or marmots, or dormice reflect the general pattern for all hibernators.

There are few studies in which the precise functional status of a particular endocrine gland is measured at the time an animal first becomes torpid. Chronic intubation, a technique which permits blood sampling at the time animals are entering dormancy or have been dormant for known periods, has long been available and telemetry makes it possible to measure the thermal history of particular animals. However, little progress has been made in combining these procedures to assay hormonal levels at the first onset of torpor. Even if blood levels are measured at the precise time an animal enters hibernation, one is only measuring the equilibrium between glandular secretion and tissue uptake. If the gland were to stop secreting, while the tissue continued to utilize the hormone, blood levels would progressively fall and definitive conclusions regarding endocrine "shut-down" could be reached.

Several factors can combine to keep the hormone level unchanged or to raise it even though the gland may secrete less because of neuroendocrine inhibition (or lack of stimulation) or lowered body temperature. There may be more binding because of an increase in the level of binding proteins and/or an increase in hormone-protein affinity due to decreased temperature. Circulation to the tissues may be reduced. Finally, low body temperature may lower metabolism of the hormone.

Chronic intubation is a procedure which does not disturb hibernation. In one of the few studies using this procedure, the exact thermal history of an animal in hibernation was determined at the time blood was sampled. Kastner et al. [19] used this technique for blood sampling to examine the renin-

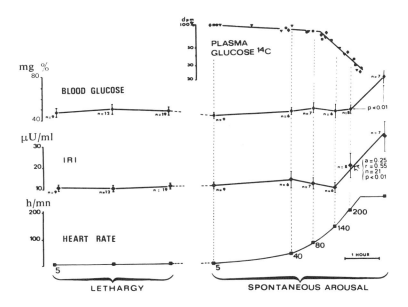

Fig. 1. Blood glucose, insulin levels and heart rate of hedgehogs (*Erinaceus europaeus*) in hibernation (lethargy) and during arousal. Hibernation bouts lasted eight to ten days. Parameters measured were taken on the first day, fourth to fifth days and eighth to tenth days. Glucose ^{14}C was injected at least 24 hours prior to arousal [11].

angiotensin-aldosterone system in the hibernating woodchuck. They concluded that the rising level of aldosterone during torpor was a consequence of low blood pressure. These responses resulted in increased plasma renin levels which, in turn, stimulated the adrenal cortex to secrete more aldosterone to increase the level of plasma potassium. Hoo-Paris et al. [11] provided specific information about the function of the insulin pancreas and glucose regulation during hibernation and arousal of the hedgehog, *Erinaceus europaeus* (Figs. 1 and 2). They used chronic intubation and continuous monitoring of heart rate and

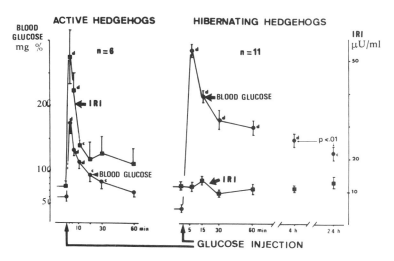

Fig. 2. Blood glucose and insulin levels of active and hibernating hedgehogs in response to the injection of 300 mg/Kg of glucose at time zero (denoted by arrows). The hibernators were studied during the middle of the hibernation bout [11].

made simultaneous measurements of subcutaneous temperature. There was a correlation between cardiac frequency and subcutaneous temperature, for example: at 20°-25°C heart rates were 140-200/min and at 15°-20°C heart rates were 80-104/min. Furthermore the data showed that insulin secretion by the pancreas does not occur until the body temperature of the arousing hedgehog reaches approximately 25°C. Thus, the insulin pancreas becomes active each time there is an episode of spontaneous arousal during the hibernation season. This system is particularly useful because it is easy to measure the endpoint of the hormonal action, viz., decrease in blood glucose, and it does not involve the complexity of neuroendocrine control.

The purpose of this review is to examine in detail the studies of the thyroid gland and its relationship to hibernation.

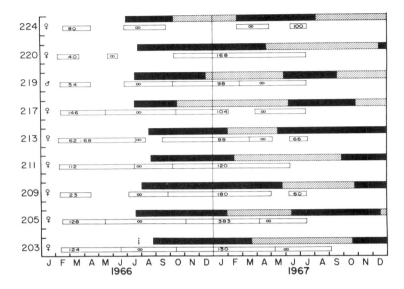

Fig. 3. The biological half-life (T 1/2) of thyroidal ^{125}I was measured over a two-year period in the desert ground squirrel, *Citellus* (*Spermophilus*) *tereticaudus*. Blackened rectangles (horizontal bars) represent periods when individual animals (vertical axis) were spontaneously torpid at night in the animal room, T_a about 22°C. Open rectangles with the enclosed numbers are the T 1/2s [15].

EXPERIMENTAL RESULTS

Many studies of the thyroid gland of ground squirrels and marmots indicate that a reduction in thyroid activity precedes hibernation [10, 18, 24, 25]. Yet the period when the thyroid gland of the desert ground squirrel, *Spermophilus tereticaudus* fails to release radioiodine is during the summer. This indicates that the gland is inactive and often precedes episodes of nightly torpor by several months [15]. The periods of thyroid inactivity, denoted by the infinity symbol in Figs. 3 and 4, occurred during the summer. In 1966, torpor did not begin until late summer or early fall in one group and fall in the other group.

38

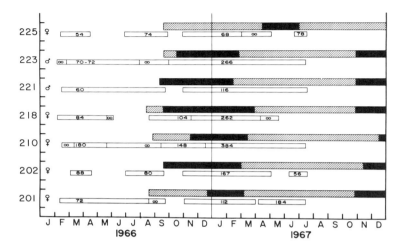

Fig. 4. Individual representatives (vertical axis) of *Citellus* (*Spermophilus*) *tereticaudus* in which both the biological half-life (T 1/2) of thyroidal 125I and spontaneous nightly torpor (blackened rectangles) were measured over a two-year period [15].

This pattern was repeated in 1967, but at an earlier date in one group and a later date in the other group. These results probably reflect desynchronization of circannual rhythms. This observation could mean that thyroid inactivity is a necessary prelude to torpor, but it is difficult to understand why it must occur as early as three to four months prior to the onset of dormancy.

During the season of nightly torpor, daily doses of L-thyroxine sufficient to double the basal (or standard) metabolic rate (Fig. 5) did not terminate nightly torpor. (Note the large number of open circles representing body temperatures below 30°C). Nightly torpor at room temperature is not, of course, the same as the level of thermal depression observed in hibernation. This suggests that thyroxine does not prevent a drop in body temperature, though body temperature does not appear to be as low at the higher cumulative doses of L-thyroxine. It also indicates that animals thought to be normothermic because they are active

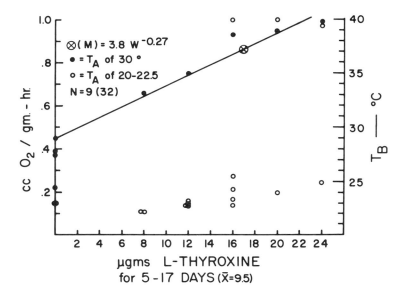

Fig. 5. The basal metabolic rate (closed circles) and body temperature (open circles) of *Citellus (Spermophilus) tereticaudus* given doses of L-thyroxine in amounts shown as the cumulative dose. Nightly torpor did not occur at an ambient temperature of 20°-22.5°C in three out of 13 measurements from nine animals. Standard metabolism was measured during the day at T_a 30°C. The predicted or expected level of standard metabolism was calculated from the equation: MR = 3.8 G 0.73.

during the day may have lowered body temperatures during a part of any 24-hour period.

In every species of ground squirrel exclusive of the nonhibernators we [14], as well as others [10], observe the thyroid gland exhibiting reduced activity beginning with summer and extending into spring, though fall and winter activity tends to be a little higher. The woodchuck, *Marmota monax* (in some considerations a large ground squirrel), exhibits a similar seasonal cycle [24]. The lowest level of total serum T_3 and T_4 occurred during the summer while the fall concentrations increased to a level 65-75% of that measured in the spring (Table 1).

The 13-lined ground squirrel exhibits the typical pattern of decreasing thyroid secretory activity as winter approaches (Fig. 6). However, it is not possible to induce thyroid secretion in nonhibernating ground squirrels during the winter by cooling the hypothalamus, a procedure known to stimulate thyroid secretion in rats [21].

TABLE 1

BLOOD T3 AND T4 LEVELS OF HIBERNATORS AT DIFFERENT SEASONS

Species	Summer	Fall	Winter	Spring	Reference
			T_4 (µg/dl)		
Erinaceus europaeus	1.2 ± 3.4	---	0.18 ± 0.7	---	[1]
Marmota monax	2.3 ± 1.0	3.2 ± 1.0	4.0 ± 0.5 (H)	5.4 ± 0.6	[25]
Spermophilus richardsoni	7.6 ± 1.2	11.3 ± 2.2	11.2 ± 1.9	---	[7]
Ursus americanus	---	2.9	1.5 ± 0.5 (H?)	2.5 ± 0.4	[2]
			T_3 (ng/dl)		
Marmota monax	45 ± 27	130 ± 12	437 ± 32 (H)	202 ± 22	[25]
Spermophilus richardsoni	90 ± 10	250 ± 15	237 ± 19	---	[7]
Ursus americanus	---	94.0	46 ± 10 (H?)	73 ± 9 83 ± 8	[2]

(H) = in hibernation
(H?) = there is some question whether bears are in hibernation or prolonged lethargy

Demeniex and Henderson [7] reported that Richardson's ground squirrel, *Spermophilus richardsoni*, exhibits its highest levels of T_3 and T_4 during the winter when it hibernates. It goes from a summer T_3 level of about 90 ng/dl to a winter level of about 550 ng/dl and a summer T_4 level of about 1.0 µg/dl to a winter level of about 6.0 µg/dl (Figs. 7 and 8). Richardson's ground squirrels kept at 18°C did not hibernate, but also exhibited a winter rise in T_3 and T_4 (Table 1), though there was less amplitude in the seasonal change than among those animals which hibernated. Demeniex and Henderson [7] interpret their data to indicate that the pituitary-thyroid axis of this species of ground squirrel is active during the winter. Their conclusions are diametrically opposite to those of Hulbert and Hudson [18] who were unable to activate the pituitary-thyroid hypothalamus in the 13-lined ground squirrel with hypothalamic chilling. Despite the T_b increases there were no increases in the loss of radioiodine from the thyroid gland (Fig. 9). Using this same radioiodine release technique, we observed a winter activity of the thyroid glands (Fig. 10) in this species, but

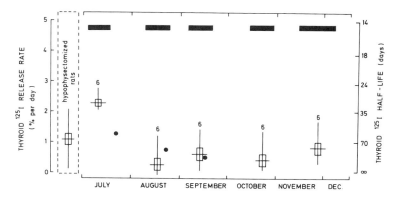

Fig. 6. The rate of ^{125}I release from the neck region of *Spermophilus (Citellus) tridecemlineatus* from July to December, compared with a hypophysectomized rat and similar data (closed circles) observed in an unpublished study (Hudson). The horizontal line represents the mean; the vertical line is the range, and the rectangles are ± s.e.m. [18].

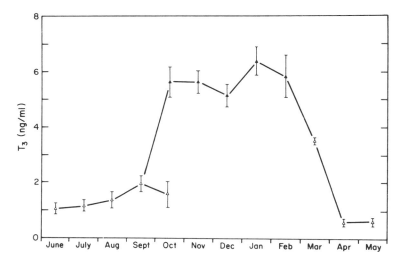

Fig. 7. Mean (± s.e.m.) monthly serum levels of total T_3 measured by radioimmunoassay in *Spermophilus richardsoni* at the time animals were in hibernation (closed triangles) or were active (normothermic) in the field (open triangles) [7].

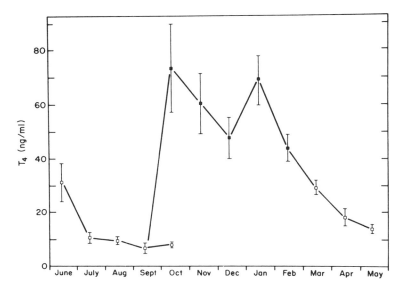

Fig. 8. Mean (+ s.e.m.) monthly serum levels of total T4 measured by radio-immunoassay in *Spermophilus richardsoni* at the time animals were in hibernation (closed squares) or were active (normothermic) in the field (open squares) [7].

were unable to induce hibernation at that time. This led us to believe that we had inadvertently misphased their circannual rhythms [14].

Seasonal cycles of thyroid secretory activity in ground squirrels are indisputable. At present it appears that these are circannual rhythms, but whether the significant change is a winter increase or a summer decrease is far less certain.

If thyroid inactivity must preceed hibernation, it would be expected that this should occur in all species of hibernators. This is not the case. Chipmunks, which belong to the same family as the ground squirrels and marmots (Sciuridae), have active thyroid glands during their hibernation season and give no indication that the gland is ever shut off [13]. Their thyroid secretory levels during the winter, a time when animals were intermittently torpid in darkness at T_a 7.5°C and after a period when the photoperiod and temperature were progressively reduced, was only slightly less than nontorpid chipmunks kept at 8°C when they were not hibernating and were on a long (14:10) photoperiod (Figs. 11 and 12). The total T4 level in the serum was almost as high in torpid animals as it was in those which were normothermic (Fig. 13). Thus, the chipmunk *Tamias striatus*, which is considered to be closely related to the group

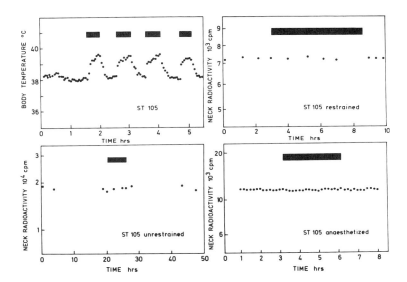

Fig. 9. The body temperature (rectal) of a *Citellus* (*Spermophilus*) *richardsoni* in which the hypothalamus has been chilled (black rectangles) by thermodal perfusion (upper left). ^{125}I release from the neck during hypothalamic chilling (blackened rectangles) with a restrained, unrestrained and anaesthetized ground squirrel was measured by placing animals over a surface crystal connected with a scaler. Closed circles are the measured levels of radioactivity (10^3 cpm) both before, during, and after chilling [18].

ancestral to ground squirrels [12], does not exhibit a seasonal shutting off of the thyroid gland. Significantly, chipmunks are inhabitants of the open forest, where solar radiation is less intense than the open grasslands or deserts where ground squirrels live.

Chipmunks are not the only species of hibernators which have active thyroids during hibernation. Tashima [23] demonstrated that the golden hamster, *Mesocricetus auratus* kept in the cold and in total darkness for 30 days, conditions sufficient to induce hibernation, had active thyroids at the time they should have hibernated. The fact that Tashima neither observed hibernation in her

44

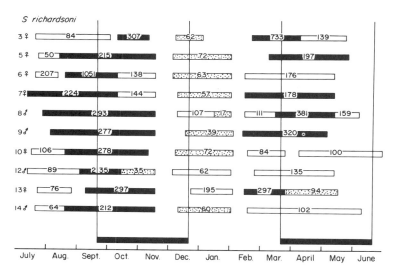

Fig. 10. The biological half-life (T 1/2) of ^{125}I (numbers in the rectangles) measured in *Spermophilus richardsoni* for an entire year. Periods for an entire year are depicted by horizontal bars, including: periods of least activity (slowest release) = black, periods of greatest activity = open, while periods of intermediate levels of activity = stippled. The vertical lines connected with blackened bars represent fall and spring even though animals were kept at a constant room temperature with a 12:12 photoperiod [14].

animals nor examined blood samples from animals in hibernation, lessened the impact of the study.

In our studies of the Turkish hamster, *Mesocricetus brandti*, the thyroid gland is active not only up to the time of hibernation, but also during hibernation. Animals kept on a short photoperiod are supposed to be reproductively inactive [8], a prerequisite for hibernation. Our animals exhibited dormancy within a few days after exposure to T_a 5°C. Their blood levels of total T_3, T_4 and TSH increased markedly within hours after being changed from room temperature to 5°C (Figs. 14, 15, and 16). Also, T_3 and TSH remained high during the period when animals were periodically dormant (Figs. 17 and 18). Interestingly,

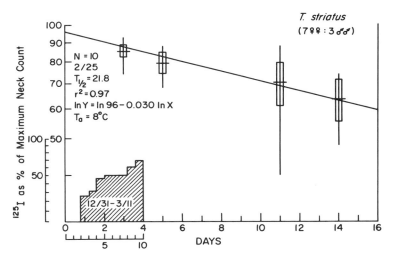

Fig. 11. Release of ^{125}I from the neck region of chipmunks is shown. Most animals had been torpid for varying periods of time during the experiment (lower hatched figure showing the percent time each had been observed in torpor at least once each day). Two of the ten animals were never torpid and one was torpid 70% of the time. Horizontal lines are means; vertical lines are ranges, and rectangles are ± 2 s.e.m. The line was fitted by least squares without day zero being defined as 100% so that the line intercepts the ordinate at about 96% [13].

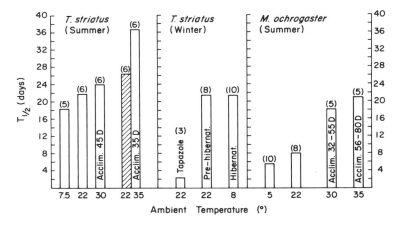

Fig. 12. The biological half-life (T 1/2) of ^{125}I of chipmunks, *Tamias striatus*, both normothermic and torpid, and voles, *Microtus ochrogaster* (a non-hibernator equivalent in size to chipmunks) exposed to a variety of ambient temperatures (°C), in some cases sufficiently long to have become acclimated.

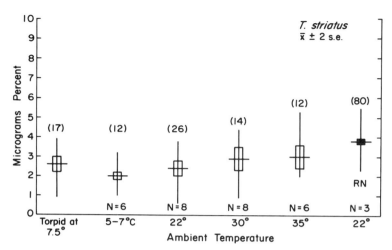

Fig. 13. Serum levels of total T4 in torpid chipmunks and chipmunks kept at a variety of air temperatures. Serum from a white rat (blackened rectangle) was used as a control. Horizontal lines represent means, vertical lines the range, and the rectangle is ± s.e.m. Numbers in brackets are the total number of times N individuals were measured [13].

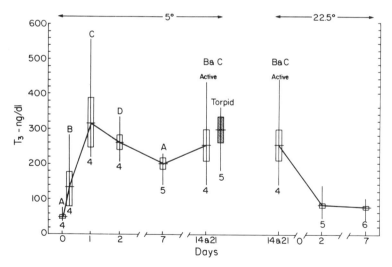

Fig. 14. Total serum T3 of the Turkish hamster, *Mesocricetus brandti* prior to (day zero) and during exposure to T_a of 5°C, at which time all but one animal was repeatedly torpid (Fig. 17). Upon exposure to room temperature (T_a 22.5°C),

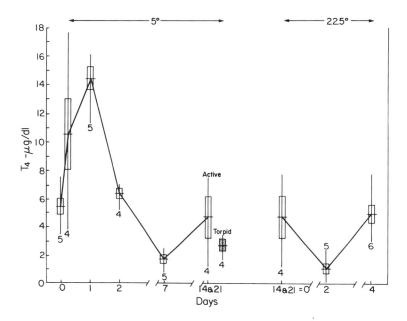

Fig. 15. Total serum T4 of the Turkish hamster, *Mesocricetus brandti*, measured before (Day 0) and after exposure to T_a 5°C; all but one of the 20 animals exhibited torpor at some time (Fig. 17). Serum samples were from the same animals in Fig. 14. Groups B and C were combined to separate those in torpor from those normothermic at the time of sampling. T4 levels in torpid animals were just barely below those of nontorpid animals. With a sample size of four, the difference is not statistically significant. Symbols are described in the legend for Fig. 14.

while the T3 level remains high, the T4 level declines. This may mean that accompanying enhanced thyroid secretion during cold exposure there must be increased activity of deiodinating enzymes so that a higher proportion of T4 is converted to T3 [4]. It might appear that hamsters do not show an increased rate of thyroid hormone secretion, measured by the release of ^{125}I from the neck region, since cold exposure results in only a transient increase in the release rate (Fig. 19). This contrasts with chipmunks (Fig. 12) which show a signif-

[Fig. 14 legend continued] T3 level declines to within 20 ng/dl of original value. Values are given as means (horizontal lines), ranges (vertical lines), and ± 2 s.e.m. (rectangles). A total of 20 animals was used, and sampled in groups of five (alphabetical letters). Groups B and C were measured on Day 14 and Day 21, and results were combined to separate those in torpor from those normothermic at the time of sampling. There is no significant difference in the serum levels of T3 of active and torpid hamsters.

48

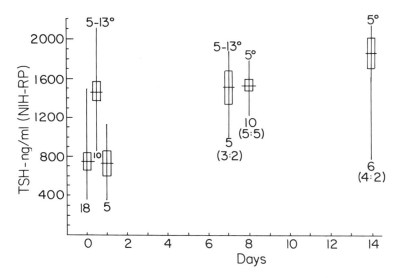

Fig. 16. Thyroid stimulating hormone (TSH) measured using NIH rat antigen and antibodies. In one chamber the temperature cycled between 5° and 13°C, thus it was necessary to repeat the experiment to ascertain TSH levels on Day 14. Note the similarity of values at Day 7 between the first group kept in T_a 5°-13°C chamber and second group with constant T_a at 5°C. The mean is given as horizontal lines, range as vertical lines and ± 2 s.e.m. as rectangles. TSH values are given as NIH reference preparation standards, which can be converted to Rat mU/ml when multiplied by 0.22.

icant decrease in the T 1/2 during cold exposure. However, Bauman et al. [3] report an increase in the thyroid secretory rate of the golden hamster, *Mesocricetus auratus*, using procedures which precluded thyroidal reuse of iodide released from thyroxine metabolism. This indicates that iodide trapping by the thyroid is probably much more effective in hamsters than in chipmunks.

The work of Canguilhem [6] indicates that the European hamster, *Cricetus cricetus*, will not hibernate without functional thyroid glands. This might also be the case in the Turkish and golden hamsters, although there do not appear to be any reports that thyroidectomies prevent hibernation in these species.

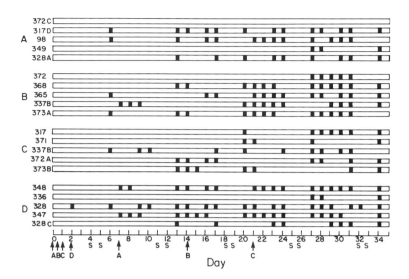

Fig. 17. Episodes of torpor (blackened rectangles) in each *M. brandti* measured at the beginning of each day, exclusive of Saturday (S) and Sunday (S). Arrows denote times each group was sampled and retro-orbital blood was obtained for assay of total T_3 and T_4. Frequency of torpor was greater during the fourth week, when animals were not disturbed, than during the first week.

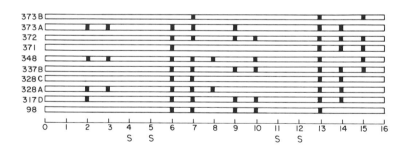

Fig. 18. Episodes of torpor (blackened rectangles) in each *M. brandti*. Blood was taken retro-orbitally to measure TSH. Only those animals observed in torpor at some time during the experiment are shown. No observations were made on either Saturday (S) or Sunday (S).

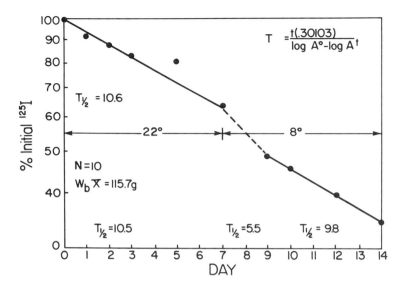

Fig. 19. Release of radioiodine (^{125}I) from the neck region of ten *M. brandti* at room temperature followed by exposure to 8°C. There is a marked increase in release rate within the first 48 hours after exposure to the low temperature, followed by a return to the room temperature release rate. Lines were fitted by eye; T 1/2 was calculated from the equation in the upper right corner and shown below each line.

DISCUSSION

Research in this area has produced a great deal of facts and deductions which must be evaluated. Why are the blood levels of T_3 and T_4 in ground squirrels and marmots higher during the winter and spring, a period encompassing the hibernation season, than during the summer? It may be that the high blood levels of winter and spring are due to the greater binding at the low body temperature of hibernation, as well as lower peripheral utilization, so that T_3 and T_4 accumulate in the blood. Young et al. [25] propose that periodic arousal of hibernators leads to an increase in free T_3 and T_4 in plasma. Since binding is reduced with elevation of body temperature, this provides an essential hormone without requiring increased thyroid activity. Curiously, these authors dismiss a 30% increase in serum protein as a significant factor in the enhancement of winter binding.

The mode of action of T_3 and T_4 requires entry into cells to bind with nuclear receptors. Without knowing the relative competition between nuclear receptors and plasma proteins as a function of temperature, it is difficult to assess quantitatively the importance of blood levels of free T_3 and T_4. It is not even clear from the literature whether albumins or globulins are the important binding proteins [24, 26].

Why is it necessary to assume that the significant seasonal change is an increased blood level of T_3 and T_4 during the winter and spring? Is it not just as reasonable to assume that the important change is the lowered levels of the summer? Generally ignored in the various studies of thyroid function in ground squirrels (including marmots) is their strict diurnality and propensity for basking, as well as distribution in habitats where solar radiation may make the dissipation of endogenous heat, particularly during exercise, difficult. Thus, it would be advantageous to have a relatively inactive thyroid during the summer as a mechanism for reducing endogenous heat production. This would explain why the lowest levels of blood T_3 and T_4 are observed during the summer rather than the fall [7, 25].

It has been well established that ground squirrels have annual rhythms which are endogenous, no doubt related in part to their fossorial habitat. Environmental cues may not be available to phase their life cycles with appropriate ambient conditions. Such rhythms could just as easily be reflected in endocrine secretions as they are in the frequently measured parameters of body weight and temperature. Is it not reasonable to expect that endocrine changes might occur while the animals are in their hibernation season as preparation for the active period? Hamsters, on the other hand, are cued by the photoperiod, as is evidenced by the role of the pineal gland [22]. Further, ground squirrels store energy as fat in preparation for hibernation, while hamsters store food and periodically arouse to eat. Perhaps thyroid inactivity is essential for fat deposition. Certainly it is clear that not all species of hibernators require thyroid secretory quiessence as preparation for hibernation. Clearly, ground squirrels and marmots are different from hamsters. There is evidence also that other endocrine glands and regulatory functions differ among hibernators. Recent reports describe differences in carbohydrate metabolism in Scuridae and the Cricetidae (Musacchia and Deavers, this volume).

Why then, is the thyroid of ground squirrels and marmots inactive during hibernation? It has been proposed that thyroid inactivity is a mechanism for increasing membrane fluidity, which allows animals to tolerate low body temperatures [17]. If hamsters do not have this seasonal thyroid inactivity, how are we to rationalize thyroid quiessence as a general physiological adaptation for hibernation? There is at least one other hibernator, the hedgehog,

Erinaceus europaeus, which also exhibits an increase in membrane fluidity associated with reduction in thyroid activity [1]. Similar studies need to be done on hamsters to determine if there is a change in membrane fluidity, and if so, whether or not it is associated with any reduction in thyroid activity. Certainly there is not a dramatic decrease such as observed in ground squirrels and hedgehogs.

CONCLUSIONS

Just as we have come to appreciate the significant differences between species of hibernators with respect to their thermoregulatory performances, we must learn to recognize that the endocrine systems of different species of hibernators also differ. For reasons not yet understood, hamster thyroid secretory activity is entirely different than that of ground squirrels. Perhaps this is related to their relative differences in degree of dependence on endogenous annual rhythms.

Many investigators have been misled by generalizing from a single species or even a few species. All too often we ignore the ecological context within which different kinds of hibernators evolved and live. During the day, active ground squirrels must contend with periods of high air temperature and intense solar radiation, whereas nocturnal hamsters avoid these conditions. Ground squirrels undergo a fattening process to prepare for hibernation, whereas chipmunks and hamsters rely on food caches. It is conceivable that thyroid quiessence is important not only for the facilitation of fat deposition, but also to minimize the quantity of endogenous heat which must be lost at times when body-to-air temperature gradients are small.

If the hypothesis of thyroid inactivity as a prerequisite for increasing membrane fluidity is to be widely applicable, there must be significant differences in the depth of body temperatures under the natural conditions for hibernation of various species. Resolution of these contradictions will require the study of a variety of different kinds of hibernators. Thus, another good case for the value of comparative physiology can be made.

ACKNOWLEDGMENTS

The National Institute of Arthritis, Metabolism and Digestive Diseases (NIAMDD), Rat Pituitary Hormone Distribution Program, and Dr. A. F. Parlow generously supplied the reagents used to measure TSH in the hamster. All of the original research reported here, as well as the time made available for writing, were supported by an NSF Grant PCM-7914629.

REFERENCES

1. Augee, M. L., Raison, J. K. and Hulbert, A. J. (1979) Seasonal changes in membrane lipid transitions and thyroid function in the hedgehog. Am. J. Physiol., 236, E 589-593.

2. Azizi, F., Mannix, J. W., Howard, D. and Nelson, R. A. (1979) Effect of winter sleep on pituitary-thyroid axis in American black bear. Am. J. Physiol., 237, E 227-230.

3. Bauman, T. R., Anderson, R. R. and Turner, C. W. (1968) Thyroid hormone secretion rates and food consumption of the hamster (Mesocricetus auratus) at 25.5° and 4.5°C. Genl. Comp. Endocrinol., 10, 92-98.

4. Bernal, J. and Refetoff, S. (1977) The action of thyroid hormone. Clin. Endocrinol., 6, 227-249.

5. Bigelow, W. G. and Sidlofsky, S. (1961) Hormones in hypothermia. Br. Med. Bull., 17, 56-60.

6. Canguilhem, B. (1970) Effets de la radio thyroidectomie et des injections d'hormone thyroidienne sur l'entree en hibernation du Hamster d'Europe (Cricetus cricetus). Seances Soc. Biol., 164, 1366-1367.

7. Demeneix, B. A. and Henderson, N. E. (1978) Serum T_4 and T_3 in active and torpid ground squirrels, Spermophilus richardsoni. Genl. Comp. Endocrinol., 35, 77-85.

8. Hall, V. and Goldman, B. (1980) Effects of gonadal steroid hormones on hibernation in the Turkish hamster (Mesocricetus brandti). J. Comp. Physiol., 135(2) B, 107-114.

9. Hoffman, R. A. (1964) Terrestrial animals in cold: Hibernators, in Handbook of Physiology, Dill, D. B., ed. Am. Physiol. Soc., Washington, D.C., Sect 4, pp. 379-403.

10. Hoffman, R. A. and Zarrow, M. X. (1958) A comparison of seasonal changes and the effect of cold on the thyroid gland of the male rat and ground squirrel (Citellus tridecemlineatus). Acta Endocrinol., 27, 77-84.

11. Hoo-Paris, R., Castex, CH. and Sutter, B. CH., J. (1978) Plasma glucose and insulin in the hibernating hedgehog. Diabete and Metabolism (Paris), 4, 13-18.

12. Hudson, J. W. (1978) Shallow, daily torpor: A thermoregulatory adaptation, in Strategies in Cold: Natural Torpidity and Thermogenesis, Wang, L. C. H. and Hudson, J. W., eds. Academic Press, New York, pp. 67-108.

13. Hudson, J. W. (1980) The thyroid gland and temperature regulation in the prairie vole, Microtus ochrogaster and the chipmunk, Tamias striatus. Comp. Biochem. Physiol., 65A, 173-179.

14. Hudson, J. W. and Deavers, D. R. (1976) Thyroid function and basal metabolism in the ground squirrels, Ammospermophilus leucurus and Spermophilus spp. Physiol. Zool., 49, 425-444.

15. Hudson, J. W. and Wang, L. C. (1969) Thyroid function in desert ground squirrels, in Physiological Systems in Semiarid Environments, Hoff, C. C., and Riedesel, M. L., eds. Univ. New Mexico Press, Albuquerque, N.M., pp. 17-33.

16. Hudson, J. W. and Wang, L. C. H. (1979) Hibernation: Endocrinologic aspects. Ann. Rev. Physiol., 41, 287-303.

17. Hulbert, A. J. (1978) The thyroid hormones: A thesis concerning their action. J. Theor. Biol., 73, 81-100.

18. Hulbert, A. J. and Hudson, J. W. (1976) Thyroid function in a hibernator, *Spermophilus tridecemlineatus*. Am. J. Physiol., 230, 1138-1143.

19. Kastner, P. R., Zatzman, M. L., South, F. E. and Johnson, J. A. (1978) Renin-angiotensin-aldosterone system of the hibernating marmot. Am. J. Physiol., 234(5), R178-R182.

20. Kayser, C. (1961) The Physiology of Natural Hibernation. Pergamon Press, New York, Oxford, London, Paris, 325 pp.

21. Reichlin, S. (1964) Function of the hypothalamus in regulation of pituitary-thyroid activity, in Brain-Thyroid Relationships with Special References to Thyroid Disorders, Cameron, M. P. and O'Connor, M., eds. Ciba Foundation Study Group No. 18, Little, Brown & Company, Boston.

22. Reiter, R. J. (1973) Endocrine rhythms associated with pineal gland function, in Biological Rhythms and Endocrine Function, Hedlund, L. W., Franz, J. M. and Kenny, A. D., eds. Plenum Press, New York, London, pp. 43-78.

23. Tashima, L. S. (1965) The effects of cold exposure and hibernation on the thyroidal activity of *Mesocricetus auratus*. Gen. Comp. Endocrinol., 5, 267-277.

24. Wenberg, G. M. and Holland, J. C. (1973) The circannual variations of thyroid activity in the woodchuck (*Marmota monax*). Comp. Biochem. Physiol., 44A, 775-780.

25. Young, R. A., Danforth, E., Jr., Vagenakis, A. G., Krupp, P. P., Frink, R. and Sims, E. A. H. (1979) Seasonal variation and the influence of body temperature on plasma concentrations and binding of thyroxine and triiodothyronine in the woodchuck. Endocrinology, 104, 996-999.

26. Yousef, M. K. and Johnson, H. D. (1975) Thyroid activity in desert rodents: A mechanism for lowered metabolic rate. Am. J. Physiol., 229, 427-431.

Published 1981 by Elsevier North Holland, Inc.
Musacchia and Jansky, eds.
Survival in the Cold
Hibernation and Other Adaptations

THE REGULATION OF CARBOHYDRATE METABOLISM IN HIBERNATORS

X. J. MUSACCHIA AND D. R. DEAVERS[1]
Department of Physiology and Biophysics, University of Louisville, Louisville,
Kentucky 40292, U.S.A. [1]New address: Physiology/Pharmacology Discipline, College of Osteopathic Medicine and Surgery, 3200 Grand Avenue, Des Moines, Iowa
50312, U.S.A.

INTRODUCTION

The biological phenomenon of hibernation is a complex series of events controlled by regulatory mechanisms which can be recognized and separately investigated.   Also, there is little question that these mechanisms are interrelated.
The events, between and including the time an animal enters into and arouses
from hibernation, can be classified as: (1) entry into a state of hibernation
when the animal passes from a physiologically active state with body temperatures (T_b) about 38°C to physiologically depressed torpid states with T_b often
4°-7°C, (2) duration of a bout of hibernation, i.e., the time during which the
subject is in a state of depressed metabolism, with body temperatures close to
ambient temperatures (T_a), and (3) arousal from hibernation when inherent
regulatory signals stimulate the animal to arouse to a fully active physiological state with body temperatures of about 38°C.

These states refer only to individual bouts of hibernation and do not
elucidate the variations of duration of hibernation bouts during the hibernal
season [36, 38].   Another aspect which is often relevant to investigations in
the physiology of hibernation is the circannual preparation for hibernation.

It is well known that some hibernators prepare for winter torpor by increasing the deposition of body fat and with hypertrophy of brown adipose tissue
(BAT). These tissues become the primary energy source during the hibernating
season and during the arousal processes.   The visual detection of fat accumulation is not distinct over short daily intervals but is obvious if one were to
examine an animal at intervals of several weeks during late summer and fall.
Some of the best examples of such hibernators are ground squirrels.   It is not
unusual for an adult, fully mature ground squirrel to gain 50-80% in body weight
during August and September. It is noteworthy that species which show considerable weight gain, such as ground squirrels, generally do not store food.   In
contrast, species such as the hamster store considerable quantities of food and
show limited weight gain in the fall; in fact, they lose body weight after cold
exposure prior to hibernation [16].

Numerous investigators have made us aware that entry into hibernation involves a variety of specific events. For example, there have been studies of central nervous system (CNS) responses which could be interpreted as test entries into depressed metabolic states [35]. More than a decade ago, the relationship between sleep and entry into hibernation had been researched by South et al. [33] and more recently Heller [12] reviewed the subject of thermoregulatory similarities between sleep and hibernation (this volume, by Heller et al.). Evidence has been presented for the accumulation of low molecular weight peptide compound(s) which promotes entry into hibernation [4, 5, 34]. The subject of hibernation inducing compounds is reviewed by Swan (this volume) and Oeltgen and Spurrier (this volume) present a substantive chemical characterization of the hibernation induction trigger (HIT).

There have been substantial investigations of hormonal and metabolic regulation in hibernators which have focused on thermogenic mediators such as the thyroid hormones and catecholamines. By comparison, there have been relatively few studies concerned with the regulation of carbohydrate metabolism. Often these have centered on measurements of blood and tissue carbohydrate composition during unspecified stages of hibernation. In our opinion one reason for this lack of experimental interest is that few investigators have been able to correlate the variation in tissue carbohydrates with specific stages in hibernation. In addition, the growing realization that there are considerable differences in regulatory mechanisms among hibernators has made it difficult to develop a unifying hypothesis concerning the regulation of carbohydrate metabolism in hibernation.

Many efforts have been made to characterize hibernation with common blood chemistry characteristics [1, 6, 7, 11, 23, 24, 36], but it is fair to say that the present state of knowledge reveals that many lacunae exist and few conclusive deductions can be made.

In recent years glucocorticoids have been shown to be important in mechanisms relatable to carbohydrate homeostasis and thermogenesis in hibernation [6, 23] and in thermoregulatory functions in nonhibernators [19, 20]. It has been recognized for some time that adrenalectomy precludes an animal's entry into hibernation [14, 25, 26, 37]. Also, few if any adrenalectomized hamsters or ground squirrels enter hibernation unless they are treated with cortisone or deoxycorticosterone (DOC) or a small fraction of the adrenal cortex is left in sito [14, 25, 37].

Our recent studies have focused on several aspects of carbohydrate and tissue carbohydrates in hibernators. One aspect has been a comparison of blood and tissue carbohydrates in Syrian hamsters (*Mesocricetus auratus*) and Turkish hamsters (*Mesocricetus brandti)* during cold exposure and continuous hibernation.

Our approach has been to use hamsters which have been in continuous hibernation, without arousal, for periods of one to six days. In this way it is possible to examine features which are relatable to the duration of a bout of hibernation. In the past it has been more common to compare parameters in hibernating versus nonhibernating subjects. Unfortunately, that approach tends to consist of random measurements and has led to some confusing interpretations.

RESULTS AND DISCUSSION

Comparative aspects of blood glucose and liver glycogen during hibernation: There is no clear picture of carbohydrate metabolism in hibernating species. The two parameters for routine measurements are blood glucose levels and tissue glycogen contents. Metablic pathways, in and of themselves, are critical in assessments of glucose availability and glycogen synthesis and degradation. However, specific approaches become appropriate only after there is an understanding of the general features of tissue carbohydrates and hibernation. For this reason, our research has focused on changes of blood glucose and tissue glycogen in the hamster during sequential days of hibernation. We believe that the best way to understand the role of carbohydrates in the physiological adjustments of an animal in hibernation is to make day-to-day measurements. To provide support for this contention, we have made an evaluation of data for ground squirrels, *Citellus* (*Spermophilus*), from reports in the literature, and an assessment for hamsters, *Mesocricetus,* chiefly with recent data from our laboratory and earlier data from Lyman's laboratory.

Gluconeogenesis and glycogenolysis are well recognized sources of glucose in the hibernating animal [3]. Blood glucose is relatively unchanged during hibernation in the hamster [18] and was thought to be so in the Arctic ground squirrel [24]. However, Hannon and Vaughan [11] were of the opinion that marked hypoglycemia takes place in hibernating Arctic ground squirrels. Such differences have been difficult to resolve until recently when Galster and Morrison [9] reported variations in blood glucose levels that depended on the time during the hibernation bout when the blood sample was collected. For example, blood glucose levels are highest during the early stages of a hibernating bout (159 mg%) and lowest during the late stages (59 mg%) (Table 1).

TABLE 1

BLOOD GLUCOSE CONCENTRATIONS OF HIBERNATING AND
NONHIBERNATING GROUND SQUIRRELS AND HAMSTERS

Species	Non-Hibernating	Hibernating	Source	
	(mg%)	(mg%)		
Citellus undulatus	153	148	Musacchia & Wilber	[24]
Citellus undulatus	156	56	Hannon & Vaughan	[11]
Citellus undulatus	125	59 (late)a	Galster & Morrison	[9]
Citellus undulatus	125	159 (early)b	Galster & Morrison	[9]
Citellus lateralis	175	125	Twente & Twente	[36]
Citellus tridecemlineatus	111	87	Burlington & Klain	[3]
Citellus tridecemlineatus	150-145 (Pre)(Post)	152	Agid et al.	[2]
Citellus armatus	128	121 (1-3 days)	Cranford (personal communication)c	
Mesocricetus auratus	113	126	Lyman & Leduc	[18]
Mesocricetus auratus	98	100	Present study	
Mesocricetus brandti	85	90	Present study	

aLate within a bout of hibernation
bEarly within a bout of hibernation
cMean bout length = 11.7 days

A summary of comparative data from various authors is presented in Table 1.
The Musacchia and Wilber [24] report dealt with Arctic ground squirrels that
were dug out of the ground (by the first author) during hibernation under field
conditions. No record of the length of time each animal had been torpid in a
hibernation bout was available. The animals were deeply torpid with rectal
temperatures (T_{re}) uniformly low (1° to 4°C) and the blood glucose values might
reflect early or midstages in a bout of hibernation. Hannon and Vaughan [11]
maintain that there were no differences between animals which had hibernated for
a very short time and those which hibernated for a prolonged period (> 7 days).
Also, they reported that blood glucose levels in newly awakened animals (> 1
day) were no different than those which had not hibernated. Their findings
indicated a marked fall in blood glucose during hibernation (Table 1). Their

view was that changes reflected the physiological state, i.e., hibernation versus nonhibernation, and were not necessarily related to entry or arousal.

Additional studies by Twente and Twente [36] using *C. lateralis* and Burlington and Klain [3] using *C. tridecemlineatus* showed significant reduction in blood glucose during hibernation (Table 1). Another recent study using *C. armatus* showed a moderate decrease during the first three days of a bout (Cranford, personal communication). Thus, if we accept the Galster and Morrison [9] contention that blood glucose levels are labile and reflect the length of time in hibernation, it appears that our [24] Arctic ground squirrels may have been sacrificed early in a hibernating bout. In general, the weight of the findings in the Sciurids strongly favors the contention that blood glucose is decreased during a bout of hibernation.

The question of carbohydrate utilization during hibernation is perhaps more reasonably approached on the basis of species differences. It appears that hibernating Cricetids, such as hamsters, maintain blood glucose levels at ranges comparable to values in normothermic active hamsters (Table 1). Both the Syrian hamster and the Turkish hamster show essentially unchanged levels of blood glucose during four to five and six days of hibernation, respectively (Figs. 1 and 2). In these studies blood was collected from animals that had hibernated continuously during normal bouts of hibernation. These data suggest that the maintained blood glucose levels are regulated during hibernation. This hypothesis is further strengthened when one considers that in the hypothermic hamster, i.e., the animal which is experimentally induced into a state of depressed metabolism [6, 22], there is progressive hypoglycemia and death in about 24 hours. Blood glucose levels (mgs/100 ml) in hypothermia can be controlled by glucose infusion if the maintenance level in the blood is 40-45 mgs% of the normal value (about 100 mg%). In these animals, there is an increase in survival from one to four days [29]. It is important to note that four days in hypothermia (T_{re} 7°C) are about the same as a typical bout of hibernation in the Syrian hamster.

The above explanations of carbohydrate utilization and blood glucose regulation are supported by measurements of liver glycogen. Liver glycogen, an important source of circulating glucose, is often measured concomitantly with blood glucose. In this instance the results of various authors are not as contradictory. With the exception of Hannon and Vaughn (Table 2), we note that in representatives of both Sciurids and Cricetids there is a marked reduction in liver glycogen during the course of hibernation. Data from Lyman and Leduc [18], using *M. auratus*, indicate a substantial reduction and our results with

SYRIAN HAMSTERS

N = normothermic, Tre 37°C
RT = room temp., Ta 24° C
CE = cold exposed, Ta 7°C

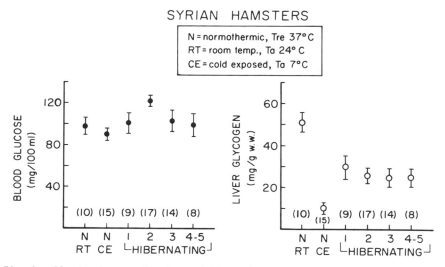

Fig. 1. Blood glucose levels and liver glycogen contents in Syrian hamsters during four to five consecutive days of hibernation, room temperature (RT) controls and cold exposed (CE) normothermic subjects (one to three weeks); T_{re} = rectal temperature, T_a = ambient temperature. Data are expressed as M and SEM and numbers of animals are in parentheses.

TURKISH HAMSTERS

N = normothermic, Tre 37°C
RT = room temp.,Ta 24°C
CE = cold exposed, Ta 7°C

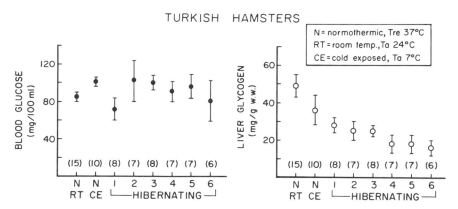

Fig. 2. Blood glucose levels and liver glycogen contents in Turkish hamsters during four to five consecutive days of hibernation, room temperature (RT) controls and cold exposed (CE) normothermic subjects (one to three weeks); T_{re} = rectal temperature, T_a = ambient temperature. Data are expressed as M and SEM and numbers of animals are in parentheses.

the same species show a fall of about 50%. Our interpretation of the fall in liver glycogen content is that the preparation for hibernation, the prolonged cold exposure prior to hibernation, and the entry into hibernation are energetically costly and glycogen is the readily available metabolic fuel.

TABLE 2
LIVER GLYCOGEN CONTENT OF HIBERNATING AND
NONHIBERNATING GROUND SQUIRRELS AND HAMSTERS

Species	Non-Hibernating	Hibernating	Source	
	(mg/gm wet weight)			
Citellus undulatus	41.0	39.0	Hannon & Vaughn	[11]
Citellus undulatus	14.4	6.1 (late)[a] 14.5 (early)[b]	Galster & Morrison	[9]
Citellus tridecemlineatus	39.5	27.2	Burlington & Klain	[3]
Mesocricetus auratus	36.8	31.5	Lyman & Leduc	[18]
Mesocricetus auratus	51.0 25.0 (CE)[c]	25.0	Present study	
Mesocricetus brandti	49.0	22.0	Present study	

[a]Late within a bout of hibernation
[b]Early
[c](CE) Prolonged cold exposure, a prerequisite for hibernation in hamsters.

One of the major differences between ground squirrels and hamsters is the fact that the former readily enter hibernation, often within several hours of cold exposure. To understand the role of liver glycogen in the hamster, it is essential to include evaluations of cold exposed animals. Our experiments showed that with prolonged cold exposure, a prerequisite for hibernation in the Syrian hamster, liver glycogen levels were reduced (Table 2). Although it would be tempting to draw a sharp distinction between ground squirrels and hamsters, it is important to note that the Turkish hamster enters hibernation more readily than the Syrian hamster. In this respect, Turkish hamsters bear some resemblance to ground squirrels. In one experiment [7] both species of hamsters were exposed to low ambient temperature (T_a) 7°C. Many Turkish hamsters hibernated within the first day, whereas the earliest entry into hibernation among the Syrian hamsters was about a week; most subjects entered after a few weeks.

```
N  = normothermic, Tre 37°C
RT = room temp., Ta 24°C
CE = cold exposed, Ta 7°C
```

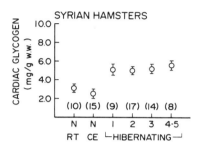

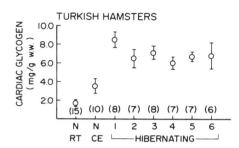

Fig. 3. Cardiac glycogen contents in Syrian and Turkish hamsters during four to five and six consecutive days of hibernation, respectively; room temperature (RT) controls and cold exposed (CE) normothermic subjects (one to three weeks). Data are expressed as M and SEM and numbers of animals are in parentheses.

Our records also showed that, during any given day after a week of cold exposure, 80% of the Turkish hamsters were in hibernation and only 25% of the Syrian hamsters were in hibernation [7].

Cardiac and skeletal muscle glycogen are often considered an important source of carbohydrate for local energy requirements rather than a more generalized storage depot. Hannon and Vaughn [11] noted marked increases in heart and skeletal muscle glycogen in hibernating *C. undulatus*. Lyman and Leduc [18] reported a slight increase in muscle glycogen during hibernation in the Syrian hamster and a marked increase in the heart glycogen content. Our data show a marked increase in the cardiac glycogen in both Syrian and Turkish hamsters during hibernation (Fig 3). These results are comparable with those of Lyman and Leduc.

Skeletal muscle glycogen in Syrian hamsters and in Turkish hamsters show considerable variation during hibernation (Fig. 4). Such changes are suggestive of cyclical turnover of muscle glycogen. This area is relatively unexplored and requires additional investigation.

Use of experimental hypothermia as a model for hibernation: In recent years there has been growing evidence that carbohydrate metabolism during hibernation is regulated via distinct hormonal pathways. Much of the work leading to these ideas is in an incomplete state and much of the evidence is indirect. We have

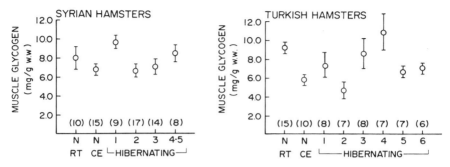

Fig. 4. Muscle glycogen contents in Syrian and Turkish hamsters during four to five and six consecutive days of hibernation, respectively; room temperature (RT) controls and cold exposed (CE) normotheric subjects (one to three weeks). Data are expressed as M and SEM and numbers of animals are in parentheses.

explored known data and examined the relative significance of these data with respect to several concepts which have gained favor. For example, our laboratory has provided evidence which indicates that glucocorticoids play a role in the maintenance in and arousal from hibernation in the Syrian hamster. This conclusion was reached after several experiments were conducted using both hibernating and hypothermic animals. Much of the work from our laboratory has been summarized in Fig. 5.

The relationship between hibernating and hypothermic hamsters evolved some years ago when it was noted that hamsters in hibernation maintained blood glucose levels at about the same or higher levels than those in normothermic animals (open triangles, Fig. 5). Similar data had been reported earlier by Lyman and Leduc [18]. On the other hand, hamsters that were induced into hypothermia showed marked hypoglycemia with levels reaching 31.5 mg% and 5.5 mg% depending upon the length of time in hypothermia at T_{re} 7°C (closed circles, Fig. 5). Since blood glucose levels were so markedly reduced in hypothermia, experiments were designed to test the need for glucose in sustaining the hypothermic metabolically depressed subject. In brief, it was found that when glucose was infused or injected into hypothermic Syrian hamsters to an amount which provided a maintenance level of about 45 mg% blood glucose, survival of

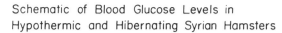

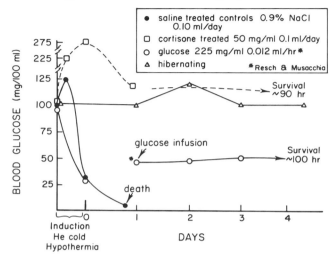

Fig. 5. A composite of several investigations dealing with blood glucose changes during hypothermia and hibernation. The open circles include the work of Resch and Musacchia [29]; closed circles include the work of Prewitt et al. [27] and Deavers and Musacchia [6]; the open squares include the work of Musacchia and Deavers [23] and Deavers and Musacchia [6]; the open triangles summarize data from the present investigation and from Deavers and Musacchia [7].

the animal in hypothermia T_{re} 7°C was sustained for a period as long as four days (open circles, Fig. 5). This period is comparable to bouts of hibernation in the Syrian hamster; they seldom exceed four to five days. Hypothermic hamsters may be unique in their ability to utilize glucose during hypothermia since other species are generally unable to do so when glucose is given intravenously [1, 25].

It has been apparent, however, that in order to utilize the hypothermic subject as a model for hibernation some regulating agent would need to be identified. Knowing that the adrenal is critical to hibernation, our attention focused on the glucocorticoids. Using cortisone, as a representative glucocorticoid hormonal regulator, it was learned that both glycemia and survival qualitites could be markedly improved (open squares, Fig. 5). Furthermore, glucocorticoid treated hamsters showed survival periods which are similar to the duration of normal bouts of hibernation. Resch [30] found that in the hamster insulin retains some of its functional integrity during hypothermia. His report adds to the growing evidence that carbohydrate metabolism is maintained, albeit at low level.

Glucocorticoids and thermogenesis: Despite the fact that hamsters can be sustained in hypothermia using exogenous glucose, they do not spontaneously awaken as do hibernating subjects. This absence of arousal capacity raised questions concerning the role for regulatory mechanisms of glucose production and in the arousal phenomen. The role of glucocorticoids in thermogenesis were recently reviewed [6] and recent studies [7, 23] employing cold exposure and hypothermia have been useful in elucidating mechanisms for glucocorticoids in thermogenesis. There is compelling evidence, from species unable to hibernate, that glucocorticoids are necessary and that they potentiate mechanisms needed in thermogenesis.

Maickel et al. [19, 20] found that adrenalectomized rats, maintained on saline and 5% glucose in their drinking water and kept at room temperature, had normal body temperatures, normal plasma glucose and normal free fatty acid levels. However, exposures to 4°C for three hours produced significant declines in body temperature. This reduced body temperature could not be restored by the catecholamine epinephrine, and epinephrine treatment did not significantly elevate plasma free fatty acids or glucose levels in these rats. In contrast, cortisone treatment partially restored the body temperature. Neither epine- phrine treated adrenalectomized rats nor untreated adrenalectomized rats showed visible signs of piloerection, vasoconstriction or shivering.

Maickel et al. [19, 20] also investigated the potentiating effects of glu- cocorticoids on plasma free fatty acids and glucose levels. Treatment with epinephrine or norepinephrine alone had no major effects on free fatty acid or blood glucose levels of cold exposed adrenalectomized rats. However, when these rats were pretreated with cortisone and then given either epinephrine or nor- epinephrine, plasma free fatty acid levels doubled and there were significant increases in plasma glucose levels. In this regard Shafrir and Steinberg [32] noted that adrenalectomized dogs were unable to increase plasma free fatty acids in response to epinephrine injection. However, if the dogs were maintained on cortisone, there was a significant elevation of plasma free fatty acids in response to epinephrine treatment.

There is evidence from animals capable of natural hibernation or experimental hypothermia that implicate glucocorticoids as being important in thermogenic mechanisms. In hamsters pretreated with cortisone acetate (i.p. 5 mg/day for several days) prior to hypothermic induction to T_{re} 7°C, there is a definite hyperglycemia immediately following attainment of T_{re} 7°C. This is in marked contrast to the hypoglycemia seen in untreated or saline treated hypothermic hamsters [6, 27]. Also, cortisone pretreatment prevents the depletion of liver glycogen which normally occurs during the hypothermic induction period [6].

It was noted in our earlier experiments that hypothermic hamsters do not arouse to normothermia when kept at an ambient temperature of 7°C [21, 27] even when provided with glucose intravenously [29]. However, glucocorticoid pretreatment resulted in arousal to normothermia in 50% of the hamsters which were in hypothermia [6, 23]. Thus, in addition to elevating glycemic levels, glucocorticoids appear to have a thermogenic effect. Another species capable of hibernation, the euthermic European hedgehog, has recently been shown to dramatically increase heat production following injection of Metopirone [39]. These results have been interpreted as evidence for an initial increase in 11-deoxycorticosteriods which increase nonshivering thermogenesis. Also, injection of deoxycorticosterone caused an increase in body temperature that was not limited to areas containing brown fat. It is reasoned that the thermogenic response to deoxycorticosterone appears to differ from the thermogenic response following norepinephrine treatment [41].

Hamsters which had glucocorticoid pretreatment appear to have additional benefial effect. Those hamsters that did not spontaneously arouse following regular glucocorticoid pretreatment regime showed much improved survival. Hamsters that were given an additional cortisone treatment after they were hypothermic survived for periods comparable to glucose infused hamsters [23].

In our view there are several possible mechanisms to explain the enhanced thermogenic capacity in glucocorticoid treated hypothermic hamsters, although much of our deductions are based on evidence from experiments with rats. In the rat it has been shown that glucocorticoids have a potentiating effect in enhancing mobilization of free fatty acids by catecholamines [19, 20]. Also, there is information that in vitro preparations of white fat made from adrenalectomized animals show blunted responses to lipolytic hormones [8, 10, 31]. In those experiments, addition of glucocorticoids restored the lipolytic effects. There is, therefore, compelling evidence that glucocorticoids, by their permissive potentiation of catecholamines, could provide a major energy source in the form of free fatty acids. Although not yet tested, one can hypothesize that free fatty acids are a potential source of energy for arousal in the glucocorticoid treated hypothermic hamster.

Glucocorticoids have been implicated in permissive action in the pressor effects of catecholamines [13, 15, 28]. In one instance, selective vasomotor control with peripheral vasoconstriction and vasoconstriction of the posterior corpus of the arousing hibernating animal is interpreted as preventing excessive heat loss. The utilization of free fatty acids during the arousal process is another general observation of most or all hibernating species during the arousal process. These two events, which appear to require glucocorticoids, are reasoned to be very important in allowing sufficient thermogenesis during arousal in glucocorticoid treated hypothermic hamsters.

It is possible that a more direct action of glucocorticoids or a permissive action with hormones other than catecholamines, such as is found in the European hedgehog by Wünnenberg and Merker [41], are coming into action in these hamsters. The fact that blood glucose levels are high and liver glycogen is not depleted immediately after hypothermic induction in cortisone treated animals also is thought to contribute to the mechanism of increased thermogenesis in arousal of hypothermic hamsters. Glucocorticoids apparently have both a direct involvement in promoting gluconeogenesis and a permissive involvement including the potentiation of glucagon-induced gluconeogenesis [40] for piloerection and shivering thermogenesis appears to play a role in glucocorticoid induced enhancement of thermogenesis [19, 20]. Also, it should be kept in mind that these mechanisms are not mutually exclusive and all or many could contribute to enhanced thermogenesis. These various regulatory mechanisms are summarized in Fig. 6.

Recent experiments using a modification of helium cold hypothermic induction: In our most recent experiments hypothermia is induced using the same He:O_2 (80:20) mixture but with more extreme cold exposure, a T_a -8°C. Animals that are taken from the relatively warm animal room (21°-24°C) generally become hypothermic (T_{re} 7-8°C) within four hours of exposure to He:O_2 (80:20) and T_a -8°C. When these hypothermic hamsters are transferred to a cold room T_a 6°-7°C, they survive for slightly longer than 24 hours in the hypothermic condition and are not capable of arousal. The chief advantage of this modification for the induction of hypothermia is the shortened period of induction. The overall survival response is comparable to that in our earlier experiments.

Using the modified induction system described above, a study was designed to elucidate the role of cold acclimation in enhancing thermogenesis. In short, the questions raised, were: Does an animal which had previously hibernated develop a capacity for increased thermogenesis leading to arousal from hypo-

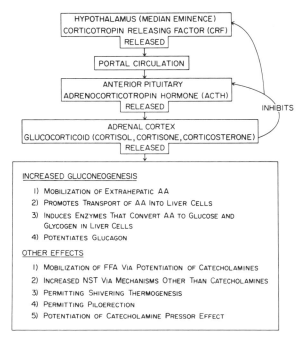

Fig. 6. A schematic presentation of known and possible regulatory functions of glucocorticoids. The control of adrenal cortex release is shown by the block diagram and arrows, including feedback inhibitory sequences. These pathways are established and known in various mammals. The large box lists known or potentially recognized effects of glucocorticoids. Many of these effects are recognized in hibernating as well as in nonhibernating species. (AA = amino acids, FFA = free fatty acids and NST = nonshivering thermogenesis).

thermia? Does an animal which had experienced cold exposure develop an improved capacity for arousal from hypothermia?

To answer these questions, animals that had been cold exposed for about three weeks (T_a 7°C) or that had experienced hibernation were transferred to an animal room (T_a 21°-24°C) for about two weeks. During the second week they were treated with cortisone or saline in a routine treatment regime [6]. Two corresponding groups of cold exposed or hibernating hamsters were also moved into the animal room for a period of about two weeks and remained untreated. Lastly, animals that had never experienced cold exposure or hibernation were either treated with cortisone or saline or remained untreated. These last groups of animals are comparable to those which had been used in previous experiments [6, 23]. All the animals in these experiments were inducted into hypothermia using the new procedure of T_a -8°C with He:O_2 (80:20) and were observed for arousal and/or survival.

All the hamsters that either had been exposed to cold or that had hibernated, whether they had received cortisone treatment, saline treatment or no treatment, aroused from hypothermia while in a cold room, T_a 6°-7°C. Animals that had never experienced cold exposure or hibernation, but had been cortisone treated, responded with arousal from hypothermia and saline treated warm room controls did not arouse from hypothermia. Thus the present results indicate that an animal's previous experience with either cold exposure or hibernation results in ability to arouse from experimental hypothermia.

In naive, i.e., noncold exposed animals, treatment with cortisone resulted in enhanced thermogenic capacity to the extent that the animals rewarmed and aroused from hypothermia. Fig. 7 compares arousal of cortisone treated, room temperature (RT) acclimated hypothermic hamsters with arousal of hamsters from natural hibernation. Arousal from hypothermia (in glucocorticoid treated or previously cold acclimated or previously hibernating hamsters) is similar to arousal from natural hibernation in that cheek pouch temperatures (T_{ch}) are always higher than rectal temperatures during the arousal process (Fig. 7). Data from arousal of naturally hibernating hamsters are from Lyman and Chatfield [17]. Had we terminated the study shortly after 100 minutes into the arousal process, as did Lyman and Chatfield, the two pairs of curves would appear even more similar. These data suggest that arousal from hypothermia, as from hibernation, is characterized by selective vasomotor control, allowing more rapid rewarming of the anterior corpus. Also, glucocorticoid pretreatment, prior cold acclimation or prior hibernation appear to enhance arousal ability with selective vasomotor control.

It is important to recall that in an earlier study [21] cold exposure resulted in increased resistance to hypothermic induction and resulted in shortened survival time. The present experiment also showed increased resistance to hypothermic induction (longer induction time), but in contrast, all animals aroused and survived. A major difference in the two experimental approaches is that in the earlier work, animals were cold exposed and placed directly into the cooling chambers and made hypothermic. In contrast, in the recent experiments, a two-week return to room temperature (T_a 22°-24°C) was interposed between the cold exposure and the hypothermic induction. This return

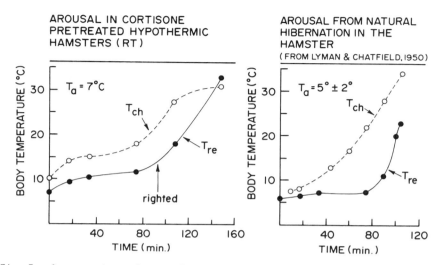

AROUSAL IN CORTISONE
PRETREATED HYPOTHERMIC
HAMSTERS (RT)

AROUSAL FROM NATURAL
HIBERNATION IN THE
HAMSTER
(FROM LYMAN & CHATFIELD, 1950)

Fig. 7. A comparison of arousal patterns in hypothermic and hibernating Syrian hamsters. Temperature recordings were made using probes in the cheek pouch (T_{ch}) and deep inside the rectum (T_{re}); ambient temperatures (T_a) also are given. At T_{res} about 13°C the hypothermic hamster is able to turn over (righted) from a supine position.

to room temperature was necessitated by the various drug treatments which were always done at T_a 22°-24°C.

To explore the difference in the arousal process between cortisone treated and saline treated hypothermic hamsters, another procedure was designed. Hypothermic hamsters, T_{re} 7°C, were transferred to a cold room at T_a 9°-14°C to stimulate arousal. Ordinarily they are maintained at T_a 7°C and untreated or saline treated controls are incapable of arousal. Measurements of cheek pouch and rectal temperatures were used in measuring arousal from hypothermia. At this ambient temperature, 9°-14°C, both the saline and cortisone treated hamsters aroused and the anterior corpus rewarmed at a faster rate than the posterior corpus.

Glucagon, another hormone which has been implicated in carbohydrate metabolism, is also known to be a gluconeogenic hormone. Since cortisone treated hamsters responded with improved levels of blood and liver carbohydrate and an enhanced thermogenic capacity leading to arousal, it was of interest to compare another gluconeogenic hormone. In preliminary studies, hamsters, taken from animal rooms T_a 21°-24°C, were treated with glucagon (i.p., 150 µg/day one day before and on the day of induction into hypothermia). Control subjects consisted of hamsters receiving an equal volume of saline (i.p., 0.9% NaCl, 0.1 ml on the same schedule). Neither saline nor glucagon treated animals were capable of arousing to normothermia in the cold room T_a 7°C. Other dosages and treatment schedules are under study.

Future evaluation will include determination of plasma glucose and plasma free fatty acids. A tentative deduction is that glucagon may not be as pervasive in its thermogenic effects. It is interesting to consider that if glucagon acts primarily to increase gluconeogenesis and glucose availability, then other substrates may be essential to promote thermogenesis to the extent necessary for arousal from hypothermia at T_a 7°C. Further experiments are necessary to confirm these deductions.

SUMMARY AND CONCLUSION

It is reasonable to conclude that significant interpretations of carbohydrate metabolism and its role in hibernation cannot be assessed by random samplings of blood and tissue constituents. Indeed, much of the literature, including one of our early contributions, must be seen as limited in value. When the pieces of this biological puzzle are assembled, the general pattern which emerges is that some hibernators such as the ground squirrel utilize carbohydrates and do not maintain blood glucose levels during a period of hibernation. On the other hand, hamsters show signs of regulated carbohydrate metabolism in which blood glucose levels are maintained during periods of hibernation. The general conclusions are that gluconeogenesis is essential to hibernation and this is best exemplified in the hamster. The hibernating ground squirrel also utilizes glucose but apparently differs from the hamster. Many questions of mechanisms about glucose utilization and synthesis remain unanswered, despite the extensive corollary studies of liver and muscle tissue glycogen.

One of the most promising areas of research dealing with the functional role of carbohydrate metabolism evolved from attempts to develop an animal model for hibernation. Using the helium-cold hypothermic hamster, we sought to define characteristics for survival in long-term states of controlled depressed metabolism. The essential requirement for glucose was established in the hypothermic hamster with T_{re} 7°C for several days.

Since hypothermic death was always paralleled with extreme hypoglycemia and since hypothermic survival was dramatically improved with glucose infusion, a hormonal regulator was sought. The ultimate decision to focus on use of a glucocorticoid provided the strongest possible evidence that gluconeogenesis not only dramatically improved survival to the point of parity with hibernation periods, but also led to the finding that cortisone treated animals spontaneously arouse from hypothermia where T_{re} is 7°C and T_a is 7°C. To date, except for our results with hypothermic hamsters, this capacity has been credited only to subjects in natural hibernation.

The rapid induction of hypothermia, using subzero temperatures in combination with helium atmospheres shows promise for enhancing our experimental protocol. Perhaps one of the most dramatic linkages between hibernation and hypothermia in the hamster is that many of the physiological parameters in hypothermia compare closely with hibernation parameters which have been investigated in over 30 years.

Much of our future research options rest within the outline presented in Fig. 6. The gluconeogenic action in the hamster is evident. One of the most promising areas is the mobilization of free fatty acids via the potentiation of catecholamines. The action of the catecholamines are numerous in regulating physiological functions in hibernation, but how about the linkage with the glucocorticoids? Their role in the arousal response from hypothermia is an entirely new observation. What are the physiological characteristics of arousal from hypothermia? For example, is there an increased level of nonshivering thermogenesis (NST) via mechanisms other than the catecholamines? These and many other questions promise a rich harvest of answers to future researchers in the fields of hibernation and hypothermia.

REFERENCES

1. Agid, R. and Ambid, L. (1969) Effects of corporeal temperature on glucose metabolism in a homeotherm, the rat, and a hibernator, the garden dormouse, in Depressed Metabolism, Musacchia, X. J. and Saunders, J. F., eds. American Elsevier, New York, pp. 119-158.

2. Agid, R., Ambid, L., Sable, R. and Sicart, R. (1978) Aspects of metabolic and endocrine changes in hibernation, in Strategies in Cold: Natural Torpidity, and Thermogenesis, Wang, L. C. H. and Hudson, J. W., eds. Academic Press, New York, pp. 499-540.

3. Burlington, R. F. and Klain, G. J. (1967) Gluconeogenesis during hibernation and arousal from hibernation. Comp. Biochem. Physiol., 22, 701-708.

4. Dawe, A. R. and Spurrier, W. A. (1969) Hibernation induced in ground squirrels by blood transfusion. Science, 163, 298-299.

5. Dawe, A. R., Spurrier, W. A. and Armour, J. A. (1970) Summer hibernation induced by cryogenically preserved blood "trigger". Science, 168, 497-498.

6. Deavers, D. R. and Musacchia, X. J. (1979) The function of glucocorticoids in thermogenesis. Fed. Proc., 38, 2177-2181.

7. Deavers, D. R. and Musacchia, X. J. (1980) Effects of cold exposure and hibernation on blood and tissue carbohydrates of Turkish and Syrian hamsters. Fed. Proc., 39, 1181.

8. Fain, J. N., Kovacev, V. P. and Scow, R. O. (1965) Effect of growth hormone and dexamethasone on lipolysis and metabolism in isolated fat cells of the rat. J. Biol. Chem., 240, 3522-3529.

9. Galster, W. and Morrison, P. R. (1975) Gluconeogenesis in Arctic ground squirrels between periods of hibernation. Am. J. Physiol., 228, 325-330.

10. Goodman, H. M. (1970) Permissive effects of hormones on lipolysis. Endocrinology, 86, 1064-1074.

11. Hannon, J. P. and Vaughan, D. A. (1961) Initial stages of intermediary glucose catabolism in the hibernator and nonhibernator. Am. J. Physiol., 201, 217-223.

12. Heller, H. C. (1979) Hibernation: Neural aspects. Ann. Rev. Physiol., 41, 305-321.

13. Kalsner, S. (1969) Mechanism of hydrocortisone potentiation of responses to epinephrine and norepinephrine in the rabbit aorta. Circ. Res., 24, 383-395.

14. Kayser, C. and Petrovic, A. (1958) Rôle du cortex surrenalien dans le mécanisme du sommeil hibernal. C. R. Soc. Biol., 152, 519-522.

15. Lepri, G. and Cristiani, R. (1964) Ability of certain adrenocortical hormones to potentiate the vasoconstrictor action of noradrenaline on the conjunctival vessels in the rabbit and in man. Brit. J. Opthalmol., 48, 205-208.

16. Lyman, C. P. (1948) The oxygen consumption and temperature regulation of hibernating hamsters. J. Exp. Zool., 109, 55-78.

17. Lyman, C. P. and Chatfield, P. O. (1956) Physiology of hibernation in mammals, in The Physiology of Induced Hypothermia. National Acad. Sci., Washington, D.C., publ. 451, 80-122.

18. Lyman, C. P. and Leduc, E. H. (1953) Changes in blood sugar and tissue glycogen in the hamster during arousal from hibernation. J. Cell. Comp. Physiol., 41, 471-488.

19. Maickel, R. P., Matussek, N., Stern, D. N. and Brodie, B. B. (1967) The sympathetic nervous system as a homeostatic mechanism. I. Absolute need for sympathetic nervous function in body temperature maintenance of cold-exposed rats. J. Pharmacol. Exp. Therap., 157, 103-110.

20. Maickel, R. P., Stern, D. N., Takabatake, E. and Brodie, B. B. (1967) The sympathetic nervous system as a homeostatic mechanism. II. Effect of adrenocortical hormones on body temperature maintenance of cold-exposed adrenalectomized rats. J. Pharmacol. Exp. Therap., 157, 111-116.

21. Musacchia, X. J. (1972) Heat and cold acclimation in helium-cold hypothermia in the hamster. Am. J. Physiol., 222, 495-498.

22. Musacchia, X. J. (1976) Helium-cold hypothermia, an approach to depressed metabolism and thermoregulation, in Regulation of Depressed Metabolism and Thermogenesis, Jansky, L. and Musacchia, X. J., eds. Charles C Thomas, Springfield, Ill., pp. 137-157.

23. Musacchia, X. J. and Deavers, D. R. (1978) Glucocorticoids and carbohydrate metabolism in hypothermic and hibernating hamsters. Effectors of thermogenesis, Girardier, L. and Seydoux, J., eds. Experientia, Suppl., 32, 247-258.

24. Musacchia, X. J. and Wilber, C. G. (1952) Studies on the biochemistry of the Arctic ground squirrel. J. Mammal., 33, 356-362.

25. Popovic, V. (1960) Endocrines in hibernation. Bull. Mus. Comp. Zool., 124, 105-130.

26. Popovic, V. and Vidovic, V. (1951) Les glandes surrenales et le sommeil hibernal. Arkh. Sci. Biol. (Belgrade), 3, 3-17.

27. Prewitt, R. L., Anderson, G. L. and Musacchia, X. J. (1972) Evidence for a metabolic limitation of survival in hypothermic hamsters. Proc. Soc. Exp. Biol. Med., 140, 1279-1283.

28. Reis, D. J. (1960) Potentiation of the vaosconstrictor action of topical norepinephrine on the human bulbar conjunctival vessels after topical application of certain adrenocortical hormones. J. Clin. Endroncrinol. Metab., 20, 446-456.

29. Resch, G. E. and Musacchia, X. J. (1976) A role for glucose in hypothermic hamsters. Am. J. Physiol., 231, 1729-1734.

30. Resch, G. E. (In press) In vivo and in vitro responses to insulin in hypothermia. Cryobiology.

31. Shafrir, E. and Kerper. S. (1964) Fatty acid esterification and release as related to the carbohydrate metabolism of adipose tissue: Effect of epinephrine, cortisol, and adrenalectomy. Arch. Biochem. Biophys., 105, 237-246.

32. Shafrir, E. and Steinberg, D. (1960) The essential role of the adrenal cortex in the response of plasma free fatty acids, cholesterol, and phospholipids to epinephrine injection. J. Clin. Invest., 39, 310-319.

33. South, F. E., Breazile, J. E., Dellman, H. D. and Epperly, A. D. (1969) Sleep, hibernation and hypothermia in the yellow-bellied marmot (M. flaviventris), in Depressed Metabolism, Musacchia, X. J. and Saunders, J. F., eds. American Elsevier, New York, pp. 277-312.

34. Spurrier, W. A., Folk, G. E., Jr. and Dawe, A. R. (1976) Induction of summer hibernation in the 13-lined ground squirrel shown by comparative serum transfusions from Artic mammals. Cryobiology, 13, 368-374.

35. Strumwasser, F. (1960) Some physiological principles governing hibernation in Citellus beecheyi. Bull. Mus. Comp. Zool., 124, 285-320.

36. Twente, J. W. and Twente, J. A. (1967) Concentrations of D-glucose in the blood of Citellus lateralis after known intervals of hibernating periods. J. Mammal., 48, 381-386.

37. Vidovic, V. and Popovic, V. (1954) Studies on the adrenal and thyroid glands of the ground squirrel during hibernation. J. Endocrinol., 11, 125-133.

38. Wang, L. C. H. (1978) Energetic and field aspects of mammalian torpor: The Richardson's ground squirrel, in Strategies in Cold: Natural Torpidity and Thermogenesis, Wang, L. C. H. and Hudson, J. W., eds. Academic Press, New York, pp. 109-145.

39. Werner, R. and Wünnenberg, W. (1980) Effect adrenocorticostatic agent, Metopirone, on thermoregulatory heat production in the European hedgehog. Pflügers Arch., 385, 25-28.

40. Wicks, W. D. (1974) The mode of action of glucocorticoids, in Biochemistry of Hormones, Rickenberg, H. V., ed. University Park Press, Baltimore, pp. 211-241.

41. Wünnenberg, W. and Merker, G. (1978) Control of non-shivering thermogenesis in a hibernator, in Effectors of Thermogenesis, Girardier, L. and Seydoux, J., eds. Experientia, Suppl., 32, 315-319.

MAXIMUM THERMOGENESIS IN HIBERNATORS: MAGNITUDES AND SEASONAL VARIATIONS

L.C. H. WANG AND BRUCE ABBOTTS
Department of Zoology, University of Alberta, Edmonton, Alberta, Canada T6G 2E9

INTRODUCTION

Maximum thermogenesis in mammals

The maximum aerobic capacity or maximum oxygen consumption ($\dot{V}_{O_2}$ max) sets the upper limit of energy which can be derived through mitochondrial respiration for work and/or thermogenesis.   The definition of such a capacity by experimental means   has been quite varied.   Hart [10] defines $\dot{V}_{O_2}$ max as the highest measurable rate or peak metabolic effort that can be sustained for a short time during severe cold exposure.   Gelineo, according to Alexander [2], uses $\dot{V}_{O_2}$ max to describe the metabolic rate which develops briefly in the first stages of hypothermia.   However, the magnitude of such a rate is sometimes difficult to ascertain since it is dependent on the rate of fall of body temperature.   A sharp decline in body temperature could suppress metabolism due to the Q_{10} effect. The $\dot{V}_{O_2}$ max thus defined is higher than the summit metabolism, which is the highest metabolic rate at normal body temperature without voluntary muscular activity [2]. Alexander [2], who used varying wind speed to control the fall of body temperature in the newborn lamb at 1°C per 20 minutes, measured what he called summit metabolism which was the average oxygen consumption over this 20-minute period.   Although Alexander [2] preferred the use of summit metabolism rather than the peak metabolic effort of Hart [10], it is likely that both could be measuring $\dot{V}_{O_2}$ max if the ambient temperatures used were not too cold to overwhelm the animal's thermogenic capacity.

The $\dot{V}_{O_2}$ max has also been defined as the peak oxygen consumption which lasts only a few minutes during exhaustive exercise [28].   This method appears to be the only viable means to elicit $\dot{V}_{O_2}$ max in relatively large animals or in animals which are well insulated.   For example, to elicit $\dot{V}_{O_2}$ max in a 1500 g winter-acclimatized snowshoe hare by cold, an air temperature of -160°C would be required [6], a rather difficult experimental condition.   In smaller animals, the recently developed technique using HeO_2 (20-21% oxygen, 80-79% helium) and cold [7] has been quite useful in eliciting $\dot{V}_{O_2}$ max [34, 44].   In this method, the six to seven times higher thermoconductivity of the helium compound to that of nitrogen facilitates heat loss, resulting in elicitation of $\dot{V}_{O_2}$ max at moderately cold temperatures. For example, $\dot{V}_{O_2}$ max can be elicited in a 235 g white rat at $-3°C$ under HeO_2 but $-30°C$ is required in air [34].

The lack of general concensus on definition and methodology has created some difficulties in comparison of reported $\dot{V}_{O_2}$ max values measured by different methods. However, as indicated by the study of Rosenmann and Morrison [34], the difference in methodology affects only slightly the outcome of $\dot{V}_{O_2}$ max measurements. For example, in the white mouse, $\dot{V}_{O_2}$ max by HeO_2 was 10% higher than that by running, and in the white rat, $\dot{V}_{O_2}$ max was 7% higher by HeO_2 than that by swimming [34]. Therefore, the $\dot{V}_{O_2}$ max values presented in this review include those elicited by cold exposure as well as by forced exercise.

In addition to the problem of variations in methodology, other natural or experimental influences which could modify the magnitude of $\dot{V}_{O_2}$ max must also be considered. The most common factors are seasonal acclimatization and temperature acclimation. For instance, winter acclimatization in the red-backed vole [35] resulted in a 56% increase in $\dot{V}_{O_2}$ max over the summer value, and cold-acclimation ($5°C$) in *Peromyscus maniculatus* resulted in a 40% increase in $\dot{V}_{O_2}$ max over that following warm-acclimation ($25°C$) [15]. Similar observations have been found in the snowshoe hare [6], rat [12], 13-lined ground squirrel [32] and golden hamster [31]. The increased $\dot{V}_{O_2}$ max after winter acclimatization or cold-acclimation is due to the enhanced capability for nonshivering thermogenesis (NST), which is additive to shivering thermogenesis [18].

A central question concerning comparison of $\dot{V}_{O_2}$ max among different species is whether the magnitude of $\dot{V}_{O_2}$ max is related to body size or whether it is determined by the ecology of the species. In view of the fact that $\dot{V}_{O_2}$ max can be increased by winter acclimatization, cold-acclimation and training, the ecology of the animal and the thermal and physical history prior to determination of $\dot{V}_{O_2}$ max must be of importance. But what if these factors have been taken into consideration, i.e., conditions which are known to improve $\dot{V}_{O_2}$ max have been optimized, will there be a relationship between $\dot{V}_{O_2}$ max and body size? This is an interesting question since it is well known that the basal metabolic rate ($\dot{V}_{O_2}$ std) of homeotherms can be predicted by body size alone [20]. If $\dot{V}_{O_2}$ max can also be predicted by body size, then it is possible to predict the metabolic ratio, i.e., $\dot{V}_{O_2}$ max/$\dot{V}_{O_2}$ std, or the factorial aerobic capacity of body size. Intuitively, the metabolic ratio is likely to be higher in the larger animal than it is in the smaller animal due to the relatively higher weight-specific basal metabolic rate in the smaller animal. However, this does not appear to be universally true in the final analysis.

The earlier published data on $\dot{V}_{O_2}$ max (between 1950-1963) have been summarized by Hart [11]. These included $\dot{V}_{O_2}$ max during cold exposure with or without superimposed exercise in eight species of rodents and one lagomorph ranging in weight between 18 and 2500 g. The $\dot{V}_{O_2}$ max was 30-35 ml $(W^{0.73}.hr)^{-1}$ in animals less than 27 g to 23 ml $(W^{0.73}.hr)^{-1}$ in animals weighing 1000 and 2500 g. The metabolic ratio was between 6.0 to 9.2 (recalculated from Table 1 [11]) with the smaller animals showing the larger ratio. Pasquis et al. [28] examined the $\dot{V}_{O_2}$ max of four laboratory species, the white mouse, white rat, guinea pig and golden hamster after warm (29-30°C), or cold (6-15°C) acclimation. The oxygen consumption during exhaustive exercise was used to determine $\dot{V}_{O_2}$ max. In cold-acclimated animals, with a weight range of 30 to 1000 g, they derived the equation: $\dot{V}_{O_2}$ max (ml.min^{-1}) = 0.436 $W^{0.73}$, with W in g of body weight. The metabolic ratio was between 6.3 and 7.2 but cannot be correlated with body size. For weights over 1000 g (dogs, men, and horses), they derived from literature the equation: $\dot{V}_{O_2}$ max (ml.min^{-1}) = 0.654 $W^{0.79}$, which predicts slightly higher $\dot{V}_{O_2}$ max than that used for smaller animals. The metabolic ratio was 15.8 for men, 18.8 for dogs and 19.8 for horses, again without apparent correlation to body size. Since the exponents of both of these equations are not significantly different from the 0.75 for the prediction of the $\dot{V}_{O_2}$ std: (ml.min^{-1}) = 0.057 $W^{0.75}$, [28], it is conceivable that the metabolic ratio is relatively constant regardless of body size but may show differences in absolute values depending on the absolute size of the animal, e.g., approximately 6-7 between 30 to 1000 g, and 16-20 between 6 to 700 kg.

Using the technique of HeO_2 and cold, Rosenmann and Morrison [34] measured maximum thermogenesis in six species of 20°-24°C acclimated laboratory and wild rodents ranging from 7 to 253 g. The metabolic ratio was between 8.4-5.1. The smaller animals appear to have higher metabolic ratios but the relationship is inconsistent. Lechner [22], in an extensive survey of $\dot{V}_{O_2}$ max and $\dot{V}_{O_2}$ std among laboratory and wild species, derived the following equation from 17 species ranging in weight from 3.3 to 2540 g. These included 12 wild species from the Orders Rodentia, Insectivora and Lagomorpha: $\dot{V}_{O_2}$ max (ml.min^{-1}) = 0.499$W^{0.678}$, W in g of body weight. This equation does not include $\dot{V}_{O_2}$ max values from animals which have been under conditions that are known to improve $\dot{V}_{O_2}$ max, such as winter acclimatization, cold-acclimation, or endurance training. Thus, the $\dot{V}_{O_2}$ max values predicted by this equation are below those by the equation of Pasquis et al. [28], which was derived from cold-acclimated animals.

The $\dot{V}_{O_2}$ std surveyed by Lechner [22] from 56 species, including 50 wild species from the Orders Rodentia, Insectivora, Chiroptera, Lagomorpha,

Carnivora, Marsupialia and Primates, resulted in the equation: $\dot{V}_{O_2}$ std (ml.min^{-1}) = 0.062 W$^{0.727}$, W in g of body weight. If one takes the ratio of these two equations by Lechner, the $\dot{V}_{O_2}$ max/$\dot{V}_{O_2}$ std can be expressed as 8.28 W$^{0.049}$. This indicates that as animal size increases, the metabolic ratio becomes smaller, and vice versa.

To illustrate, the calculated ratio is 4.3 for a 700 kg horse, and 7.4 for a 10 g mouse. On the upper weight end, the calculated ratio for the horse is only 21% of that predicted by Pasquis et al. [28] and therefore does not appear to be reliable. On the lower weight end, the calculated ratio is close to what Rosenmann and Morrison [34] have observed and therefore appears to be reliable.

Finally, in a most recent study, Prothero [33] surveyed 24 species of animals, tissues (brown fat) and plants ranging in weight between 31 mg and 100 tons for $\dot{V}_{O_2}$ max. He arrived at an exponent of 0.75 for the power function and a correlation coefficient of 0.983. He interpreted this finding as a similarity in magnitude of $\dot{V}_{O_2}$ max at the cellular level in all organisms and that the relatively higher basal metabolic rates found in homeotherms are due to the maintenance of a higher fraction of their weight-specific $\dot{V}_{O_2}$ max than either the unicellular organisms or poikilotherms [33].

It is apparent from this brief review that the $\dot{V}_{O_2}$ max of mammals can be predicted by body size, similar to the prediction of $\dot{V}_{O_2}$ std. The magnitude of predicted $\dot{V}_{O_2}$ max varies, however, depending on the equation used, which in turn depends on the experimental conditions under which the equation is derived. This points to the importance in future studies of standardizing the thermal and physical histories of the animals prior to testing. The prediction of metabolic ratio by body size, on the other hand, does not appear to be reliable. Between 10 to 1000 g, the predicted ratio is 7.4 to 5.9 by using Lechner's equations [22], and the observed ratio ranges from 14.5 [35] to 6 [11]. Between 6 to 700 kg, the predicted ratio is 5.4 to 4.3, but the measured ratio is 16 to 20 [28]. The inconsistency of this prediction appears to be due to the discrepancies in choosing the proper exponents for the body size to predict $\dot{V}_{O_2}$ max in different species.

Objective for studies of maximum thermogenesis in hibernating species

A closer examination of the available studies on $\dot{V}_{O_2}$ max indicates that with the exception of the Merriam's chipmunk, *Eutamias merriami* and the golden hamster, *Mesocricetus auratus* [22], no other hibernating species have been included in such analyses. This is surprising in view of the intense heat production that is required during periodic arousal to rewarm the body within a few hours from 2°-5°C to 37°C [41].   In northern species, such as the Richardson's ground squirrel, *Spermophilus richardsoni,* spring emergence from hibernation in Alberta occurs in mid- to late March when the ground is still frozen and covered with snow.   Air temperatures as low as -25°C are not uncommon when cold fronts pass.   Because of the short growing season, breeding must take place soon after spring emergence. The animals therefore face multiple stresses of low temperature and lack of food plus high energy expenditure for thermoregulation and behavioral exhibits such as aggression and territorial defense (for the males). Thus, in the Richardson's ground squirrel, both the periodic arousals during the hibernation season and the energetic requirements after spring emergence would demand a high capacity for thermogenesis or a high $\dot{V}_{O_2}$ max.

On the other hand, adult *S. richardsoni* disappear from above ground and commence torpor (estivation) as early as July when the mean air temperature is 20°C and burrow temperature approximates 16°C [39, 41].   Torpor extends through the winter (hibernation) until emergence in March. The young of the year commence torpor typically in the second week of September, when air temperature averages 10°C and ground temperature 13°C. It is apparent that both age classes do not encounter either short durations of intense cold or long durations of moderate cold, conditions which are known to improve maximum thermogenesis in some rodents [21], prior to the onset of their hibernation season.   Thus, if there was enhancement of $\dot{V}_{O_2}$ max for needs of periodic arousal, it would have to be acquired during the hibernating season. Due to declining soil temperature as winter progresses, the burrow temperature could be as low as 0° to -2°C in January-February [41]. It is conceivable that the animal could be acclimated to this burrow temperature during the period of homeothermy between hibernation bouts.   However, since such an intertorpor homeothermy period lasts only 7-14 hours and occurs only once every few weeks [41], it is doubtful that such irregular cold challenges could lead to physiological manifestations typical of cold-acclimation.

The remaining possibilities to be considered are that there is no special change in $\dot{V}_{O_2}$ max in conjunction with hibernation and that $\dot{V}_{O_2}$ max is relatively constant all year regardless of seasonal acclimatization.   In addition, since $\dot{V}_{O_2}$ max consists of two components, shivering and NST, a change in proportion of their contributions to $\dot{V}_{O_2}$ max could also occur in accordance with seasonality and hibernation.

We have thus set out to systematically examine the seasonal variations of $\dot{V}_{O_2}$ max in *S. richardsoni*.   The questions include: (1)   Does this hibernator have higher $\dot{V}_{O_2}$ max than that of a nonhibernating species similar in weight?   (2)

Does $\dot{V}_{O_2}$ max change with season?   (3) Does $\dot{V}_{O_2}$ max change with warm (20°C) or cold (5°C) acclimation?   (4) What effects do seasonal acclimatization and temperature acclimation have on the proportional contributions of shivering and NST to $\dot{V}_{O_2}$ max?   (5) Does peak metabolic rate observed during arousal from hibernation approximate the $\dot{V}_{O_2}$ max during euthermia?

To expand the generality of our observations in answering these questions, we have conducted limited measurements of $\dot{V}_{O_2}$ max in two other hibernators, the Columbian ground squirrel, *S. columbianus*, which does not exhibit the staggered schedule of entry into hibernation among different age classes [26], and the 13-lined ground squirrel, *S. tridecemlineatus*, a more southern species but occupying a similar habitat to that of *S. richardsoni* [25]. A literature survey has been made to include peak metabolic rates during arousal from hibernation or daily torpor as a possible indicator for $\dot{V}_{O_2}$ max. The objectives of these efforts are to provide a conceptual framework on whether hibernators have superior $\dot{V}_{O_2}$ max than their nonhibernating counterparts and whether $\dot{V}_{O_2}$ max changes seasonally in preparation for hibernation.

EXPERIMENTAL PROCEDURES

Animals

S. richardsoni was trapped 15 km south of Edmonton, and *S. columbianus* was trapped on eastern slopes of the Rocky Mountains near Gorge Creek, Alberta. The 13-lined ground squirrel, *S. tridecemlineatus*, was trapped on the University of Illinois campus at Urbana and shipped to Alberta. All animals were maintained individually under 12:12 photoperiod at $22° \pm 1°C$ unless stated otherwise and fed ad libitum with standard laboratory rat chow and water. The following experimental groups were established for each species.

S. richardsoni

a. April-Adult: captured in mid-April, tested within one week

b. June-Adult: captured in mid-June, tested within one week

c. June-Young: captured in mid-June, tested within one week

d. August-Young: captured in early August, tested within one week

e. Hibernating 1977:   captured in summer, maintained at 5°C and total darkness; tested in February after having gone through many hibernation bouts

f. Hibernating 1978:   captured in summer, maintained at 5°C and total darkness; tested in October-November after at least three bouts of hibernation

g. Cold Acclimated: nonhibernating phase animals maintained at 5°C and 12:12 for a minimum of four weeks prior to testing in April

h. Warm Acclimated: nonhibernating phase animals maintained at 20° ± 1°C and 12:12 for a minimum of four weeks prior to testing in October-November

S. columbianus

a. Hibernating 1980:   maintained at 5° ± 1°C under total darkness; tested in February after many hibernation bouts

b. Cold Acclimated   maintained at 5° ± 1°C under total darkness; no hibernation occurred before or after the testing as verified by food deprivation up to ten days

c. Warm Acclimated   maintained at 22° ± 1°C and 12:12; tested in March

S. tridecemlineatus

a. Warm Acclimated   maintained at 22° ± 1°C and 12:12; some exhibited torpor at this temperature; tested in May

Testing for Thermogenesis

Estimation of nonshivering thermogenesis: Fasted nonhibernating animals and hibernating animals after disturbed arousal were cannulated in the right jugular vein with PE 10 (Clay-Adams) under sodium pentobarbital anesthesia (50 mg/kg). Isoproterenol, a β-adrenergic agonist, was infused to stimulate NST [16] at thermoneutral temperature (20°C). The body temperature of the animal was regulated between 36° and 39°C by circulation of cold (5°C) or warm (35°C) water through copper tubing beddings in the animal chamber [1]. Oxygen consumption and CO_2 production were measured by an open-flow system and integrated automatically through a computerized data acquisition system [44]. Heat production was calculated by converting oxygen consumption to calories based on R.Q. [20]. A dose-response curve was constructed by infusing 1.0, 2.5, 5, 10, and 20 ng isoproterenol-HCl per (body weight)$^{0.74}$ per minute. The highest response in oxygen consumption after infusion was taken as the maximum NST. Preinfusion metabolic response was determined when only the drug vehicle (1 mg/ml ascorbic acid in double-distilled, deionized water) was infused.

Estimation of $\dot{V}_{O_2}$ max: The HeO2 plus cold [7] method was used to elicit $\dot{V}_{O_2}$ max in all species tested. The highest heat production in a 15-minute period was used for $\dot{V}_{O_2}$ max [40]. The selection of ambient temperature depended on the species and individual differences in thermogenic capability. In general, -10° to -20°C was used for *S. tridecemlineatus*, and -20° to -45°C for *S. richardsoni* and *S. columbianus*. Body temperature was measured before and after cold exposure to ensure that mild hypothermia had occurred. The $\dot{V}_{O_2}$ max was verified when a lower ambient temperature did not elicit a higher metabolic response on weekly cold exposure schedules.

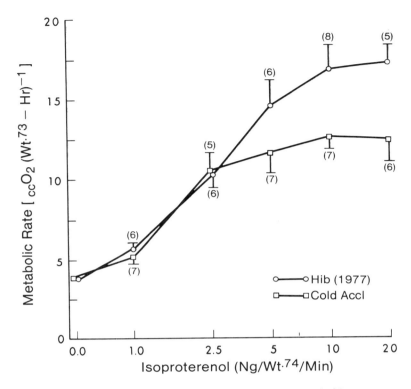

Fig. 1. Typical dose-response relationship between metabolic response and isoproterenol infusion in hibernating and cold acclimated animals are shown. Means ± 1 SE for groups [Hib (1977) and Cold Accl] are plotted. Number of measurements is shown in parentheses.

OBSERVATIONS

Metabolic response to isoproterenol in *S. richardsoni*

All groups exhibited sigmoidal dose-response relationships between heat production and the dose of isoproterenol infused. Typical results for hibernating and cold acclimated animals are shown in Fig. 1. The mean metabolic response of each group to all doses of isoproterenol is presented in Table 1. At a dose of 1 ng $(W^{0.74} min)^{-1}$, metabolic response was significantly higher than pre-infusion level in the June-Adult, Aug-Young, Hib-1978 and Cold-Accl groups (p < .05). At doses of 2.5 ng $(W^{0.74}.min)^{-1}$ or higher, metabolic responses were

TABLE 1

MEAN METABOLIC RESPONSE OF *Spermophilus richardsoni* TO DIFFERENT
DOSES OF ISOPROTERENOL

Group	Basal HP	Dose of Isoproterenol [ng(w.74-min)-1]				
		1.0	2.5	5.0	10	20
Apr-Adult	3.83+.13 (6)	4.29+.27 (6)	7.41+1.02 (6)	10.64+.80 (6)	10.51+.50 (6)	10.29+.51 (6)
June-Adult	3.46+.05 (5)	5.32+.79 (5)	7.34+1.10 (5)	9.29+.69 (5)	9.69+.86 (5)	9.42+.73 (5)
June-Young	4.27+.18 (5)	4.47+.23 (5)	6.47+.15 (5)	7.86+.40 (5)	8.11+.31 (5)	8.23+.25 (5)
Aug-Young	3.64+.21 (6)	4.16+.15 (6)	9.13+.72 (5)	10.03+.69 (6)	10.54+.60 (6)	11.14+.76 (6)
Hib-1978	3.31+.24 (7)	5.30+.56 (7)	11.37+.65 (7)	14.18+.58 (5)	14.15+.58 (7)	14.42+.32 (5)
Hib-1977	3.75+.26 (10)	5.75+.32 (6)	10.31+.78 (6)	14.55+1.68 (6)	16.85+1.46 (8)	17.26+1.09 (5)
Cold-Accl	3.83+.18 (8)	5.21+.48 (7)	10.50+1.24 (5)	11.65+1.28 (5)	12.62+.74 (7)	12.47+1.34 (6)
Warm-Accl	3.13+.26 (6)	5.37+1.24 (5)	9.64+1.07 (5)	9.82+.84 (5)	10.27+.82 (6)	10.79+.82 (5)

Means $\pm$ 1 S.E. are indicated in cc O_2 (wt.73-m-hr)-1. Number of measurements
used to determine each mean is shown in parentheses.

significantly higher than preinfusion levels in all groups (p < .05). Within
each group, the metabolic response to higher doses, 5-20 ng (W0.74 min)-1, were
not significantly different, indicating that a maximum had been attained at
doses of 5 ng or greater.

Maximum NST in *S. richardsoni*

The maximum NST was estimated by comparing metabolic responses to 10 ng
isoproterenol in each group (Fig. 2). The animals in group Hib-1977 had
significantly greater NST [79.2 cal (W0.73.hr)-1] than all other groups except
Hib-1978. Hib-1978 [66.5 cal (W0.73.hr)-1] showed greater NST than all other
groups except Cold-Accl [59.3 cal (W0.73.hr)-1]. Cold-Accl animals showed significantly higher NST than that in June-Young [38.1 cal (W0.73.hr)-1]. The
maximum NST was not significantly different in all other groups [between 40-50
cal (W0.73.hr)-1]. The only significant difference in preinfusion heat produc-

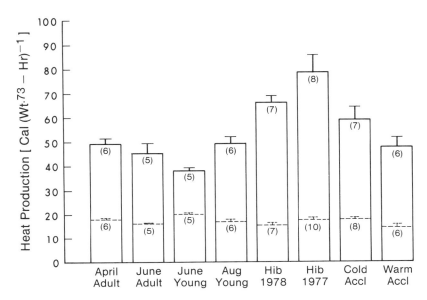

Fig. 2. Comparison of mean maximum metabolic response to isoproterenol (NST max) between groups. Histogram plots of mean responses to a dose of 10 ng $(wt^{0.74}-min)^{-1}$ ($\pm$ 1 SE) as an estimate of NST max. Dotted lines indicate mean basal heat production. Number of measurements of heat production (HP) is shown in parentheses. Significance with Duncan's multiple range test ($p < 0.01$) is shown below. Groups which are not significantly different are underlined by a common line.

Basal HP	Warm Accl	Hib 1977	June Adult	April Adult	Hib 1977	Cold Accl	Aug Young	June Young

NST max	June Young	June Adult	Warm Accl	April Adult	Aug Young	Cold Accl	Hib 1978	Hib 1977

tion was between the lowest (Warm-Accl) and the highest (June-Young) values. The latter was probably due to the young age of this group (1.5 to 2 months).

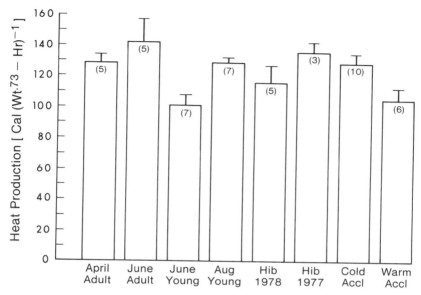

Fig. 3. Comparison of mean maximum heat production (HP max) between groups. Histogram plots mean HP max ($\pm$ 1 SE) for each group of animals. Number of measurements is shown in parentheses. Significance with Duncan's multiple range test ($p < 0.01$) is shown below. Groups which are not significantly different are underlined by a common line.

Mean Body Weight	(160) June Young	(439) Warm Accl	(391) Hib 1978	(340) Cold Accl	(276) April Adult	(382) Aug Young	(398) Hib 1977	(313) June Adult

Maximum heat production in *S. richardsoni*

The mean maximum for each group is shown in Fig. 3. The highest group (June-Adult) exhibited significantly higher heat production [142 cal $(W^{0.73}.hr)^{-1}$] than that of the lowest group [June-Young, 101 cal $(W^{0.73}.hr)^{-1}$]. However, neither of these groups was significantly different from any of the other groups (Fig. 3).

Maximum heat production in *S. Columbianus* and *S. tridecemlineatus*

In *S. columbianus*, the maximum heat production was highest in Hib-1980 [119.2 cal $(W^{0.73}.hr)^{-1}$; mean weight = 498 g, n = 6], but it was not significantly different from those of the Cold-Accl [99.9 cal $(W^{0.73}.hr)^{-1}$; mean weight = 669 g, n = 6], and Warm-Accl [91.3 cal $(W^{0.73}.hr)^{-1}$; mean weight = 702 g, n = 5] groups. In *S. tridecemlineatus* maximum heat production averaged 129.8 cal $(W^{0.73}.hr)^{-1}$ (mean weight = 203 g, n = 13). There was no consistent difference in maximum heat production between animals which were in their hibernating phase from those which were not. The individual differences in heat production were quite apparent, however, ranging from a low of 114.3 to a high of 144.5 cal $(W^{0.73}.hr)^{-1}$.

DISCUSSION

Hibernating species

In *S. richardsoni*, there is no difference in maximum NST that can be stimulated by isoproterenol between cold (5°C) or warm (20°C) acclimated groups. This is in contrast to that observed in many laboratory species such as the rat [5] and rabbit [4] which showed increases of NST after cold acclimation. The results are similar to those reported for the golden hamster acclimated to 5° and 30°C [38], for the 13-lined ground squirrel acclimated to 6° and 18°C [32] and for an unspecified species of ground squirrel acclimated to 5° and 25°C [30], all of which showed little difference in norepinephrine (NE) stimulated NST after temperature acclimation. It appears that the ability for maximum NST in the hibernators can be independent of a wide range of acclimating temperatures. However, acclimation to somewhat higher ambient temperatures such as 35°C in the hamster [3], 28°C in the 13-lined ground squirrel [32] and 36°C in an unspecified ground squirrel [29] could result in suppression of NE stimulated NST.

The lack of seasonal differences in maximum NST in *S. richardsoni* when sampled in April, June, and August, indicates that seasonal acclimatization immediately after or prior to the hibernation season has little effect on NST capability. These results are consistent with the findings of 5° and 20°C acclimation studies since both air and burrow temperatures are within this temperature range during these months. The significant increase in maximum NST during the hibernating phase of the animals (Table 1 and Fig. 2) is interesting. This is similar to that reported for another ground squirrel [30] which showed a 35.5% increase of NE stimulated NST during the hibernation season as compared to that of the 25°C acclimated animals. The corresponding increase in $S.$ *rich-ardsoni* was somewhat higher at 64% (Table 1). The mechanism via which the increase of NST is achieved during the hibernating phase of the animal is unknown. Whether this is related to any increase in the amount of brown adipose tissue or other biochemical modifications in thermogenesis remains to be investigated.

The relatively constant maximum heat production capability in $S.$ *richardsoni* (Fig. 3) regardless of seasonal acclimatization and temperature acclimation contrast with observations of other rodent species which show increased maximum thermogenesis after winter acclimatization [12, 24, 35] or cold acclimation [15, 31, 32]. Since NST and shivering thermogenesis are additive [18], we would expect an increase in maximum thermogenic capacity in conjunction with the increase of NST during the hibernating phase. The absence of a change in $\dot{V}_{O_2}$ max suggests that the maximum thermogenic capacity measured during different times of the year and after cold or warm acclimation is the maximum aerobic capacity the animal is capable of generating. If this interpretation is correct, it can be inferred that the proportional contribution of NST to this maximum is differentially increased commensurate with hibernation. Since NST is important for periodic arousal from hibernation [13, 18], it is tempting to suggest that the observed increase of NST during the hibernating season is a functional adjustment to aid the rewarming process. The increases of heat production and rate of arousal following central injection of NE [8] are consistent with this speculation since augmentation of noradrenergic outflow could result in further stimulation of NST.

It is interesting to note that the highest rates of heat production during spontaneous and disturbed arousals are between 70-93 cal $(W^{0.73}.hr)^{-1}$ [1, 41], values which are only 60-70% of the maximum rates of heat production (Fig. 3). This indicates that $\dot{V}_{O_2}$ max is not invoked during arousal in this species and that significant aerobic reserves remain in spite of the demand for intense heat production during rewarming. However, since peak metabolic rate during arousal generally occurs at a body temperature below the euthermic level in this species [41], it is possible that this discrepancy is due to a Q_{10} effect of body temperature on thermogenesis.

The results in $\dot{V}_{O_2}$ max in $S.$ *columbianus* and $S.$ *tridecemlineatus* are consistent with those described for $S.$ *richardsoni*. Thus, in these hibernators the magnitude of thermogenic capacity changes little after acclimation to 22° or 5°C; nor does it change with the onset of the hibernation season.

Comparison of $\dot{V}_{O_2}$ max between hibernating and nonhibernating species

We return to the question of whether hibernating species tend to have higher $\dot{V}_{O_2}$ max values than those of nonhibernating species. Our studies with hibernators have been compared with those in the literature (Table 2) in an attempt to answer this question. The data, taken from studies of animals in a variety of experimental conditions, include peak metabolic rates found during spontaneous or disturbed arousal from hibernation or daily torpor, as well as those stimulated by NE. These values may not be equivalent to $\dot{V}_{O_2}$ max in all the cases but they are the highest rates available for a given species. All values have been converted to express $\dot{V}_{O_2}$ max as ml min^{-1} to allow direct comparison with existing equations. It was difficult to include data from the literature where $\dot{V}_{O_2}$ max values were expressed as ml (g.hr)$^{-1}$ and weights of the animals were not given.

Fig. 4 presents a comparison of $\dot{V}_{O_2}$ max between hibernating and nonhibernating species. Only animals weighing between 10 and 1000 g are included since most hibernators fall into this weight range. The equation by Pasquis et al. [28] for cold acclimated laboratory species and the equation by Lechner [22] for 17 laboratory and wild species are used for comparison. The 95% confidence limits of Lechner's equation are shown as dashed lines. It can be seen that most of the hibernating species have $\dot{V}_{O_2}$ max which are within the 95% confidence limits of Lechner's equation. The two points below the 95% confidence limits are for induced and spontaneous arousals in the hedgehog and pocket mouse, respectively. The Eastern chipmunk, *Tamias striatis* [43] and the big brown bat, *Eptesicus fuscus* [14] have unusually high rates of oxygen consumption during arousal as well as the $\dot{V}_{O_2}$ max for *Peromyscus maniculatus* after 5°C acclimation [15]. These are the clear exceptions.

The closeness of the points from the three hibernating species studied by us, e.g., Nos. 8, 17, 21 in Fig. 4, in comparison to Pasquis' equation for cold acclimated animals, indicates that if external conditions known to enhance $\dot{V}_{O_2}$ max have been optimized, little difference in $\dot{V}_{O_2}$ max could be detected in animals of similar body size.

From these observations, it can be concluded that although periodic arousal from hibernation may impose a demand for superior thermogenesis in the hibernators, there is currently no evidence to suggest that an intrinsic difference in thermogenic capability exists between hibernators and nonhibernators. It is evident, however, that the proportional contribution of NST to $\dot{V}_{O_2}$ max increases during the hibernation season and such increases could be viewed as functional adjustments to aid in rewarming from hibernation.

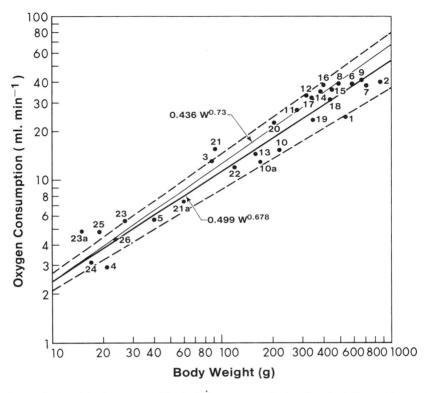

Fig. 4. Relationship between estimated $\dot{V}_{O_2}$ max and body size in hibernating and nonhibernating mammals. The heavy solid line is the equation of Lechner [22] and the dashed lines are 95% confidence limits of the exponent. The thin solid line is the equation of Pasquis et al. [28] for four cold acclimated species. Individual numbers refer to species (with corrections: 10a = 11, 21a = 23 and 23a = 26) listed in Table 2.

SUMMARY

This review of $\dot{V}_{O_2}$ max among mammals indicates that body size can be used to predict $\dot{V}_{O_2}$, as well as the basal metabolic rate. The exponent of the power function predicting $\dot{V}_{O_2}$ max ranges from 0.678 to 0.79, which is not significantly different from the 0.723 to 0.75 used to predict basal metabolic rate. Since most of the available equations are constructed using nonhibernating species, questions arise as to whether hibernating species would have higher $\dot{V}_{O_2}$ max values than those found in the nonhibernating species. Using the Richard-

TABLE 2

$\dot{V}_{O_2}$ MAX ESTIMATED BY DIFFERENT METHODS IN MAMMALS CAPABLE OF EXHIBITING HIBERNATION OR DAILY TORPOR

Animal Species No.	Wt.(g)	Conditions	ml min^{-1}	Reference	
1. *Erinaceus europaeus*	542	Induced arousal	24.6	Tähti and Soivio	[36]
2. *Erinaceus europaeus*	867	NE stimulated	40.0	Wünnenberg et al.	[45]
3. *Mesocricetus auratus*	88	Induced arousal	13.2	Lyman	[23]
4. *Perognathus californicus*	22	Spontaneous arousal	2.9	Tucker	[37]
5. *Perognathus hispidus*	40	Spontaneous arousal	5.7	Wang and Hudson	[42]
6. *Spermophilus columbianus*	592	Induced arousal	39.2	This study	
7. *Spermophilus columbianus*	701	22°C acclimated, HeO$_2$	38.7	"	
8. *Spermophilus columbianus*	497	5°C acclimated hibernating phase, HeO$_2$	39.3	"	
9. *Spermophilus columbianus*	669	5°C acclimated nonhibernating phase, HeO$_2$	40.9	"	
10. *Spermophilus lateralis*	220	Spontaneous arousal	15.3	Hammel et al.	[9]
11. *Spermophilus lateralis*	169	Spontaneous arousal	13.0	Horwitz et al.	[17]
12. *Spermophilus richardsoni*	276	April-Adult, HeO$_2$	27.7	This study	
13. *Spermophilus richardsoni*	313	June-Adult, HeO$_2$	33.4	"	
14. *Spermophilus richardsoni*	159	June-Young, HeO$_2$	14.5	"	
15. *Spermophilus richardsoni*	381	Aug-Young, HeO$_2$	35.2	"	

TABLE 2 Continued

Animal Species No.	Wt.(g)	Conditions	ml min⁻¹	Reference
16. *Spermophilus richardsoni*	447	Adult, hibernating phase, 1978, HeO₂	35.4	This Study
17. *Spermophilus richardsoni*	398	Adult, hibernating phase, 1977, HeO₂	38.0	"
18. *Spermophilus richardsoni*	340	5°C acclimated nonhibernating phase, HeO₂	32.2	"
19. *Spermophilus richardsoni*	439	20°C acclimated nonhibernating phase, HeO₂	31.3	"
20. *Spermophilus richardsoni*	437	Induced arousal	23.7	"
21. *Spermophilus tridecemlineatus*	203	22°C acclimated, HeO₂	22.7	"
22. *Tamias striatus*	92	Spontaneous arousal	15.6	Wang and Hudson [43]
23. *Eutamias minimus*	60	5°C acclimated, NE stimulated	7.2	Hayward [13]
24. *Eliomys quercinus*	120	Spontaneous arousal	12.0	Pajunen [27]
25. *Myotis myotis*	27	Induced arousal	5.6	Jansky and Mejsnar [19]
26. *Eptesicus fuscus*	15	Induced arousal	4.9	Hayward and Ball [14]
27. *Peromyscus maniculatus*	17	25°C acclimated, HeO₂	3.1	Heimer and Morrison [15]
28. *Peromyscus maniculatus*	19	5°C acclimated, HeO₂	4.8	Heimer and Morrison [15]
29. *Peromyscus leucopus*	23	Winter acclimatized, NE stimulated	4.3	Lynch [24]

$ml \ min^{-1}$ column values as shown.

son's ground squirrel, the $\dot{V}_{O_2}$ max at different times of year and after temperature acclimation was measured to elucidate the influence of seasonal acclimatization and warm and cold acclimation on the magnitude of $\dot{V}_{O_2}$ max in a hibernator. The capacity for nonshivering thermogenesis was also assessed by infusion of isoproterenol. Supplementary studies on $\dot{V}_{O_2}$ max were conducted on two other hibernating species, the Columbian ground squirrel and the 13-lined ground squirrel.

In the Richardson's ground squirrel, the $\dot{V}_{O_2}$ max is relatively constant regardless of seasonal acclimatization, temperature acclimation and hibernating status. Measurements in the Columbian and 13-lined ground squirrels yielded similar results. The maximum nonshivering thermogenic capability in the Richardson's ground squirrel, however, was significantly higher during the hibernating phase, suggesting a role in aiding rewarming from hibernation.

Comparison of $\dot{V}_{O_2}$ max from nonhibernators to those observed in the present study, as well as those from the literature for hibernating species, indicates that there is no significant difference between the two groups. In both groups, $\dot{V}_{O_2}$ max can be predicted by body size within a 95% confidence limit of the equation derived by Lechner [22] with only a few exceptions. The relatively constant $\dot{V}_{O_2}$ max found in the hibernators in all seasons suggests that conditions known to enhance $\dot{V}_{O_2}$ max in nonhibernators (winter acclimatization, cold acclimation) have already been optimized in the hibernators. The only change is the increase of proportional contribution of nonshivering thermogenesis to $\dot{V}_{O_2}$ max in conjunction with the hibernation season.

ACKNOWLEDGMENTS

Experiments reported here are supported by a Canadian National Science and Engineering Research Council Operating Grant No. A6455 to L. C. H. Wang. We thank Dr. John Willis for helping with the collection and shipment of the 13-lined ground squirrels.

REFERENCES

1. Abbotts, B. (1979) Aspects of thermogenesis in a seasonal hibernator, *Spermophilus richardsonii*. University of Alberta, Edmonton, Alberta, M.Sc. Thesis, 56 pp.

2. Alexander, G. (1962) Temperature regulation in the new-born lamb. V. Summit metabolism. Aust. J. Agric. Res., 13, 100-121.

3. Cassuto, Y. and Amit, Y. (1968) Thyroxine and norepinephrine effects on the metabolic rates of heat-acclimated hamsters. Endocrinology, 82, 17-20.

4. Cottle, W. A. (1963) Calorigenic response of cold-adapted rabbits to adrenaline and noradrenaline. Can. J. Biochem. Physiol., 41, 1334-1337.

5. Cottle, W. A. and Carlson, L. D. (1956) Regulation of heat production in cold-acclimated rats. Proc. Soc. Exp. Biol. Med., 92, 845-849.

6. Feist, D. D. and Rosenmann, M. (1975) Seasonal sympatho-adrenal and metabolic responses to cold in the Alaskan snowshoe hare (*Lepus americanus* MacFarlani). Comp. Biochem. Physiol., 51A, 449-455.

7. Fischer, B. A. and Musacchia, X. J. (1968) Responses of hamsters to He-O_2 at low and high temperatures: Induction of hypothermia. Am. J. Physiol., 215, 1130-1136.

8. Glass, J. D. and Wang, L. C. H. (1979) Effects of central injection of biogenic amines during arousal from hibernation. Am. J. Physiol., 236, R162-R167.

9. Hammel, H. T., Dawson, T. J., Abrams, R. M. and Andersen, H. T. (1968) Total calorimetric measurements on *Citellus lateralis* in hibernation. Physiol. Zool., 41, 341-357.

10. Hart, J. S. (1957) Climatic and temperature induced changes in the energetics of homeotherms. Rev. Can. Biol., 16, 133-174.

11. Hart, J. S. (1971) Rodents, in Comparative Physiology of Thermoregulation, Whittow, G. C., ed. Academic Press, New York, Vol. 3, pp. 2-150.

12. Hart, J. S. and Heroux, O. (1963) Seasonal acclimatization in wild rats (*Rattus norvegicus*). Can. J. Zool., 41, 711-716.

13. Hayward, J. S. (1971) Nonshivering thermogenesis in hibernating mammals, in Non-shivering Thermogenesis, Janský, L., ed. Swets and Zeitlinger, Amsterdam, pp. 119-134.

14. Hayward, J. S. and Ball, E. G. (1966) Quantitative aspects of brown adipose tissue thermogenesis during arousal from hibernation. Biol. Bull., 131, 94-103.

15. Heimer, W. and Morrison, P. (1978) Effects of chronic and intermittent cold exposure on metabolic capacity of *Peromyscus* and *Microtus*. Int. J. Biometeorol., 22, 129-134.

16. Horwitz, B. A. (1978) Neurohumoral regulation of non-shivering thermogenesis, in Strategies in Cold: Natural Torpidity and Thermogenesis, Wang, L. C. H. and Hudson, J. W., eds. Academic Press, New York, pp. 619-653.

17. Horwitz, B. A., Smith, R. E. and Pengelley, E. T. (1968) Estimated heat contribution of brown fat in arousing ground squirrels (*Citellus lateralis*). Am. J. Physiol., 214, 115-121.

18. Janský, L. (1973) Nonshivering thermogenesis and its thermoregulatory significance. Biol. Rev., 48, 85-132.

19. Janský, L. and Mejsnar, J. (1971) Nonshivering thermogenesis during arousal from hibernation, in Non-shivering Thermogenesis, Janský, L., ed. Swets and Zeitlinger, Amsterdam, pp. 139-145.

20. Kleiber, M. (1961) The Fire of Life. John Wiley and Sons, Inc., New York, pp. 125-201.

21. LeBlanc, J. (1978) Adaptation of man to cold, in Strategies in Cold: Natural Torpidity and Thermogenesis, Wang, L. C. H. and Hudson, J. W., eds. Academic Press, New York, p. 695-715.

22. Lechner, A. J. (1978) The scaling of maximal oxygen consumption and pulmonary dimensions in small mammals. Resp. Physiol., 34, 29-44.

23. Lyman, C. P. (1948) The oxygen consumption and temperature regulation of hibernating hamsters. J. Exp. Zool., 109, 55-78.

24. Lynch, G. R. (1978) Seasonal changes in thermogenesis, organ weights, and body composition in the white-footed mouse, Peromyscus leucopus. Oecologia (Berl.), 13, 363-376.

25. MacClintock, D. (1970) Squirrels in North America. Van Nostrand Reinhold Co., New York, pp. 37-48.

26. Michener, G. R. (1977) Effects of climatic conditions on the annual activity and hibernation cycle of Richardson's ground squirrels and Columbian ground squirrels. Can. J. Zool., 55, 693-703.

27. Pajunen, I. (1976) A comparison of oxygen consumption and respiratory quotients in Finnish and French garden dormice, Eliomys quercinus L., hibernating at 4.2 ± 0.5°C. Ann. Zool. Fennici, 13, 161-173.

28. Pasquis, P., Lacaisse, A. and Dejours, P. (1970) Maximal oxygen uptake in four species of small mammals. Resp. Physiol., 9, 298-309.

29. Petrović, V. M. and Marković-Giaja, L. (1973) A comparative study of the calorigenic action of noradrenaline in the rat and ground squirrel adapted to different temperatures. Experientia, 29, 1295-1296.

30. Petrović, V. M., Marković-Giaja, L. and Stamatović, G. (1972) Influence de la saison sur l'effect calorigenique de la noradrenaline chez le spermophile. J. Physiol. (Paris), 65, 475A.

31. Pohl, H. (1965) Temperature regulation and cold acclimation in the golden hamster. J. Appl. Physiol., 20, 405-410.

32. Pohl, H. and Hart, J. S. (1965) Temperature regulation and cold acclimation in a hibernator, Citellus tridecemlineatus. J. Appl. Physiol., 20, 398-404.

33. Prothero, J. W. (1979) Maximal oxygen consumption in various animals and plants. Comp. Biochem. Physiol., 64A, 463-466.

34. Rosenmann, M. and Morrison, P. (1974) Maximum oxygen consumption and heat loss facilitation in small homeotherms by He-O_2. Am. J. Physiol., 226, 490-495.

35. Rosenmann, M., Morrison, P. and Feist, D. (1975) Seasonal changes in the metabolic capacity of red-backed voles. Physiol. Zool., 48, 303-310.

36. Tähti, H. and Soivio, A. (1977) Respiratory and circulatory differences between induced and spontaneous arousals in hibernating hedgehogs (Erinaceus europaeus L.). Ann. Zool. Fennici, 14, 197-202.

37. Tucker, V. A. (1965) Oxygen consumption, thermal conductance and torpor in the California pocket mouse, Perognathus californicus. J. Cell. Comp. Physiol., 65, 393-404.

38. Vybiral, S. and Janský, L. (1972) Thermoregulatory significance of catecholamine thermogenesis in golden hamsters. Physiol. Bohemoslov., 21, 121-122.

39. Wang, L. C. H. (1972) Circadian body temperature of Richardson's ground squirrel under field and laboratory conditions: A comparative radiotelemetric study. Comp. Biochem. Physiol., 43A, 503-510.

40. Wang, L. C. H. (1978) Factors limiting maximum cold-induced heat production. Life Sci., 23, 2089-2098.

41. Wang, L. C. H. (1979) Time patterns and metabolic rates of natural torpor in the Richardson's ground squirrel. Can. J. Zool., 57, 149-155.

42. Wang, L. C. H. and Hudson, J. W. (1970) Some physiological aspects of temperature regulation in normothermic and torpid hispid pocket mice, *Perognathus hispidus*. Comp. Biochem. Physiol., 32, 275-293.

43. Wang, L. C. H. and Hudson, J. W. (1971) Temperature regulation in normothermic and hibernating Eastern chipmunk, *Tamias striatus*. Comp. Biochem. Physiol., 38A, 59-90.

44. Wang, L. C. H. and Peter, R. E. (1975) Metabolic and respiratory responses during Helox-induced hypothermia in the white rat. Am. J. Physiol., 229, 890-895.

45. Wünnenberg, W., Merker, G. and Bruck, K. (1974) Do corticosteroids control heat production in hibernators? Pflügers Arch., 352, 11-16.

Published 1981 by Elsevier North Holland, Inc.
Musacchia and Jansky, eds.
Survival in the Cold
Hibernation and Other Adaptations

THERMOGENESIS, BROWN FAT AND THERMOGENIN

BARBARA CANNON, JAN NEDERGAARD AND ULF SUNDIN
Wenner-Gren Institute, University of Stockholm, Norrtullsgatan 16, S-113 45 Stockholm, Sweden

Brown fat and the hibernating hamster

The hibernating hamster, with a body temperature approximately the same as the low ambient temperature must be able to rewarm itself. It can, in the course of two hours, warm itself from 5°C to 37°C and it can do this while in the cold without gaining any heat from the surroundings. It is easy to calculate* that the heat needed to rewarm the hamster demands minimal combustion of 275 mg fat. Therefore, it should be possible to look for such a decrease in body fat depots. Fig. 1 depicts the results of this type of investigation. We have looked at the interscapular deposit of brown fat in hibernating and arousing animals and followed changes in fat content.

It is evident that there was a clear decrease in the fat content in the brown fat during arousal, whereas analyzed white fat deposits did not show any such changes. The interscapular fat pad studied is only a small part of the total brown fat mass in the hamster. In a cold-adapted hamster there is as much as 1700 mg of brown fat. Provided that all brown fat deposits lose their fat to the same extent during arousal as did the interscapular fat pad, one can estimate that a total of 220 mg of fat disappears during arousal. Comparing this figure with the amount of fat theoretically needed to rewarm a hibernating hamster, calculated as 275 mg, one can conclude that the lipid loss from brown fat during arousal is nearly enough to rewarm the hamster.

Fig. 1 shows that the relative lipid content of brown fat did not change markedly during arousal, in spite of a loss of one fourth of the total amount of fat. This agrees with earlier observations [22], but it cannot be concluded from such relative determinations that there is no loss of lipid during arousal since the brown fat in hamsters, as in ground squirrels [34], actively participates in the arousal process.

---

*This value is calculated from the assumptions: a hamster body weight of 80g, a "hamster body specific heat" of 1 cal/g (4.2 J/g) and a temperature increase of 32°C (5°C → 37°C). Thus the heat needed would be about 2.6 kcal (10.7 kJ). If this heat is the result of combustion of fat, then one expects that 275 mg fat should be utilized with a calofiric value of 9.3 kcal/g (39 kJ/g).

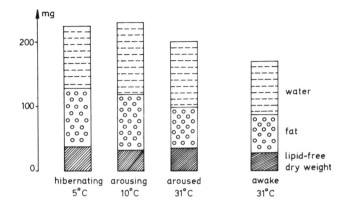

Fig. 1. The loss of fat from the interscapular deposits of brown adipose tissue
during arousal. The interscapular deposit was dissected from hibernating,
arousing, aroused and awake hamsters. Arousal was initiated by manipulation of
the hamsters. The hamsters were retained at cold 5°C during the entire arousal.
The temperatures shown were obtained on the dorsal skin surface with a thermocouple. The wet weight, the lipid-free dry weight after two extractions with
chloroform:methanol (2:1) and one extraction with heptane, and the fat weight
(mg) (the collected, evaporated extracts) were determined gravimetrically. Results shown are means from four determinations in each group, described in
[39].

It is interesting to note (Fig. 1) that hibernating animals maintain a higher
fat content than those which are not hibernating. Irrespective of whether this
fat accumulation is accomplished by the hamster before it enters hibernation in
anticipation of its later needs, or whether this extra fat is accumulated during
hibernation, when arousal is triggered, the brown fat is well equipped to
furnish the requirements of the animal for combustible fuels.

This disappearance of lipid from the brown fat during arousal does not necessarily mean that the fat is also combusted there. Indeed Hayward and Lyman [26]
demonstrated that a significant part of the heat produced during arousal in the
hamster is derived from shivering; the heating capacity of hamster brown fat
cells is not impressive at low body temperatures [46]. However, brown fat may
export substrate in the form of fatty acids to the rest of the body for combustion there, e.g., during shivering thermogenesis.

It is known from earlier work [47] that brown fat cells isolated from control
hamsters, i.e., hamsters living at ambient temperatures around 21°C, are able to
release fatty acids; this is in addition to the direct thermogenic capacity of
the cells. However, only cold-adapted hamsters are prone to enter hibernation.
We have therefore tried to study brown fat cells from cold-adapted hamsters.

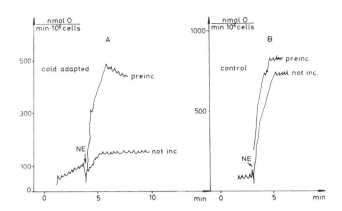

Fig. 2. Recovery of catecholamine sensitivity in brown fat cells isolated from cold adapted rats.
A. 50,000 cells from a cold-adapted rat per ml buffer.
B. 50,000 cells from a control rat per ml buffer. NE = 2 µM norepinephrine added; preinc. = cells preincubated under the conditions described in [41].

Like other investigators studying this problem [2, 3, 19, 28, 30, 52], we found that for both hamsters and rats the catecholamine sensitivity was lost when the cells were obtained from cold-adapted animals (Fig. 2A, not inc.). These cells responded poorly or not at all to norepinephrine addition. However, after a simple incubation procedure they regained catecholamine sensitivity (Fig. 2A, preinc.). Cells isolated from control animals were not affected (Fig. 2B).

A series of preincubated brown fat cell preparations obtained from cold-adapted and control hamsters have been compared [41]. The capacity for thermogenesis and fatty acid release was slight and showed statistically insignificant differences between the groups. Both had an unstimulated respiratory rate, around 50 nmol O_2 per min per million cells, and this could be increased to 400 nmol O_2 with an optimal norepinephrine concentration. Similarly, both groups had an unstimulated rate of fatty acid release of about 5 nmol per min per million cells, and this could be increased to 50 nmol with an optimal norepinephrine concentration.

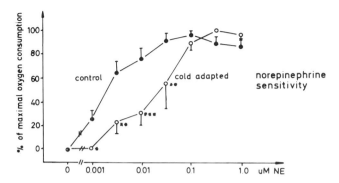

Fig. 3. Decreased norepinephrine sensitivity in brown fat cells prepared from cold-adapted hamsters. Cells were isolated, preincubated and assayed as described in [41]. Bars indicate standard errors as calculated from seven to ten determinations on five to seven preparations. From [41] by permission.

It may appear strange that cells from cold-adapted hamsters were not more thermogenic than those from control hamsters. However, brown fat cells may be in an energy-dissipating state, even in control hamsters. These hamsters obtain a palatable diet consisting of sunflower seeds, dried carrots, and maize, and may thus be in a state comparable with the "cafeteria-fed rats," studied by Rothwell and Stock [56]. It has been suggested that rats burn excess foodstuffs in brown fat, which may thus serve not only as a thermogenic but also as an antiadipogenic organ. Perhaps it can be assumed that control hamsters have a relatively large amount of brown fat, not only because they are hibernators, but also because such palatable diets induce a higher amount of brown fat, at least in rats [56].

There was one clearcut difference between the brown fat cell preparations obtained from cold-adapted and control hamsters: The sensitivity to norepinephrine was markedly lower in those preparations obtained from animals where the tissue had been in a stimulated state (Fig. 3). This could be due to differences in degradation and uptake of added norepinephrine in the different preparations. Such reactions may be important at low catecholamine concentrations [14, 20]. The difference may well be a true difference between the cells themselves and thus an expression of a "desensitization." Such a desensitization may be due to a decrease in numbers of receptors [6] or to a metabolic desensitization, e.g., an increased phosphodiesterase activity [41].

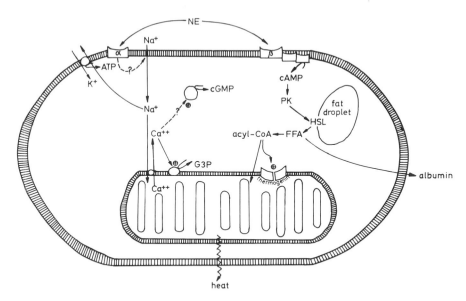

Fig. 4. Adrenergic pathways in brown fat cells. The left side of the figure illustrates the hypothetical α-adrenergic pathway discussed in the text. The right side of the figure illustrates the classical β-adrenergic pathway. PK = protein kinase, HSL = hormone-sensitive lipase, FFA = free fatty acids, G-3-P = glycerol-3-phosphate.

α- and β-adrenergic responses

That brown fat and brown fat cells respond to catecholamines as depicted in Fig. 2, i.e., with an increase in oxygen consumption--heat production [46]--, has long been acknowledged. However, as norepinephrine has almost equal affinity for α-and β-adrenergic receptors, the nature of the adrenergic response in brown fat has often been debated. General agreement exists that the major part of the thermogenic response results from stimulation of β-receptors and there is some evidence for an α-adrenergic involvement. We have suggested [48] the series of steps depicted in Fig. 4 as a hypothetical α-adrenergic pathway, and shall review some of the evidence.

First, α-adrenergic receptors seem to be present on isolated brown fat cells. At least there is evidence for high-affinity binding sites for the specific α-adrenergic antagonist dihydroergocryptine [63, 64].

Second, the membrane depolarization observed after epinephrine stimulation [24] seems to contain at least an α-adrenergic component [23], and this depolarization results from increased membrane conductance or permeability [32]. This means that ionic redistribution may take place between the medium, the interstitial fluid, and the cell after catecholamine stimulation. Indeed, a decrease in cytosolic K^+ has been measured [49]. This decrease in cytosolic K^+ is probably paralleled by an increase in cytosolic Na^+. As Na^+ promotes Ca^{++} release from brown fat mitochondria [1, 43], an increase in cytosolic Na^+ may result in an increase in cytosolic Ca^{++} from about 200 nM to 1.5 μM [40, 44, 45]. This high Ca^{++} level may have regulatory functions, either directly or through calmodulin.

One enzyme which may be regulated by cytosolic Ca^{++} is guanyl cyclase, and Skala and Knight [58] have presented evidence for an α-adrenergically stimulated increase in cyclic GMP in intact rats. Another enzyme which is regulated by cytosolic Ca^{++} is the mitochondrial glycerol-3-phosphate dehydrogenase [7]. The activity of this enzyme has been implicated in the control of the fate of fatty acids, in the form of acyl-CoA, within the cell and whether they will be combusted or esterified to triglycerides [38, 45].

Most of the processes discussed above, as possible α-adrenergic phenomena, are not very energy-demanding; e.g., "futile" Ca^{++} cycling across the mitochondrial membrane occurs under these circumstances but represents a minor energy dissipation [40]. However, an influx of Na^+ and an efflux of K^+ may result in an *indirect* stimulation of the cell membrane Na^+/K^+-ATPase, which will work in an effort to restore the ionic potential. This effect of catecholamines should be distinguished from the direct effects on the Na^+/K^+-ATPase discussed by Horwitz and Eaton [33]. The indirect stimulation of the Na^+/K^+-ATPase may lead to a significant although minor thermogenesis. Calculations [39] led to the conclusion that perhaps 10% of the total oxygen consumption during thermogenesis could result from this process. This value agrees with results from microcalorimetric experiments [13] where the effect of the Na^+/K^+-ATPase inhibitor ouabain was demonstrated to be approximately a 10% decrease in heat output. Further, in experiments with isolated cells, Mohell et al. [38] have demonstrated /that 15-30% of the maximal oxygen consumption of brown fat cells may result from stimulation of α-receptors. This conclusion arose from competitive experiments with α- and β-adrenergic agonists and antagonists. \

\ It must be stressed that the absolute dominant part of the thermogenic response of brown fat cells results from stimulation of the β-adrenergic pathway (Fig. 4), and that α-adrenergic effects may be accessory but not necessary components in the response.\ The β-receptors of isolated brown fat cells have been well characterized [65, 66] and increases in cyclic AMP level [51] and activation of protein kinase [57] have been demonstrated. The stimulation of the hormone-sensitive lipase results in the liberation of fatty acids from the triglyceride droplets. Some of these fatty acids become activated to acyl-CoAs and, in this form, are ready for degradation and combustion. It is possible that a certain fraction of the acyl-CoAs are partially degraded in the peroxisomes [35, 42], but the main portion is combusted in the mitochondria. The control of this combustion is a unique feature of brown fat mitochondria.

Thermogenin and thermogenesis

To allow foodstuffs to be combusted in an unlimited way, and without harnessing energy for the synthesis and maintenance of the cellular machinery, is a goal which is in striking contrast to the general, energy-saving concept of most living cells. It is therefore not surprising that brown fat mitochondria are endowed with a protein specifically created to allow energy dissipation to occur. For this protein we suggest the name *thermogenin*. Thermogenin is, by a somewhat cumbersome definition, probably the brown-fat-specific, GDP-(purine nucleotide-) binding polypeptide of molecular weight of 32,000 which is (at least a part of) the protonophoric--hydroxophoric (OH-) and halide conducting--complex found in the inner membrane of brown fat mitochondria, at least when the animal is in a thermogenic state. As thermogenin reveals itself in a series of ways (Table 1), a multitude of working names has been used to designate it, such as the "GDP-binding protein," the "uncoupling protein," the "32k-protein," and the "protonophore of brown fat mitochondria." It is now clear that all of these working names describe different biochemical properties of the same protein and we reasoned that a unifying name will facilitate discussion.

TABLE 1

PRESENCE OF THERMOGENIN IN BROWN FAT MITOCHONDRIA

Revealed by:	
(a)	high levels of atractyloside-insensitive purine nucleotide (GDP, ADP) binding
(b)	high rates of halide (Cl^-, Br^-) permeability
(c)	high rates of H^+ (OH-) conductance, i.e., partial uncoupling
(d)	high amounts of a polypeptide with a molecular weight 32,000

It is now possible to isolate this protein, either in low yields by using a GDP-affinity column [54] or by taking advantage of the similarities of thermogenin with the mitochondrial adenine nucleotide translocase. By using a

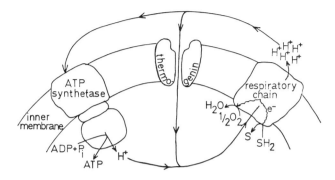

Fig. 5. Short-circuiting of the proton current by thermogenin in brown adipose tissue mitochondria. The oxidation of substrate (SH_2) is accompanied by a pumping out of protons which establishes a proton gradient over the mitochondrial inner membrane. When this gradient becomes sufficiently high, respiration ceases. Existence of the thermogenin channel allows proton reentry to occur more easily than through the ATP-synthetase path. From [9] by permission.

hydroxyapatite method similar to the one developed for adenine nucleotide translocase, Lin and Klingenberg [37] were able to obtain thermogenin in high yield. After isolation thermogenin appears to be a dimer [36], but it remains to be demonstrated that the isolated thermogenin, when incorporated into membranes (vesicles), has the properties of the native protein.

Fig. 5 presents thermogenin in its native situation in the inner membrane of brown fat mitochondria. The existence of this hole for H^+ (OH^-) allows respiration to proceed unhindered by the build up of the proton gradient [50] which controls respiration and limits heat production in all other mitochondria.

One way to measure the amount of thermogenin is to study specific binding of GDP to brown fat mitochondria (Table 1). The aim of such experiments has been mainly to correlate thermogenin and thermogenesis. Fig. 6A depicts the change in concentration of thermogenin, measured as nmol GDP bound/mg mitochondrial protein, in brown fat mitochondria during adaptation of rats to cold. During

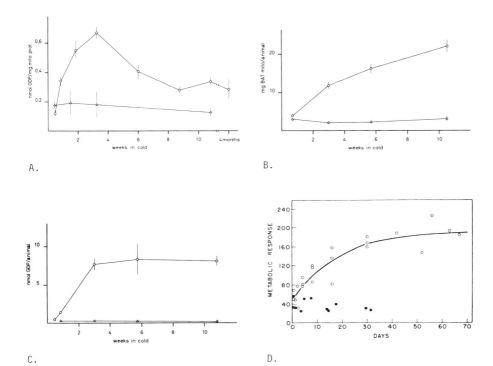

A.

B.

C.

D.

Fig. 6. Correlation between thermogenin and thermogenesis:
A. Thermogenin concentration in brown fat mitochondria.
B. Total yield of brown fat mitochondria per animal.
C. Total amount of thermogenin per animal. Open circles are mitochondria from rats at 5° C; open triangles are mitochondria from rats at 21°C.
D. The development of nonshivering thermogenesis during adaptation to cold. Open circles are rats at 6°C; closed circles are rats at 26°C.
Figs. 6A-C are from [60] by permission, and Fig. 6D from [16], by permission.

TABLE 2

CORRELATIONS BETWEEN PHYSIOLOGICAL STATE WITH NONSHIVERING
THERMOGENESIS AND THERMOGENIN

Physiological state	Nonshivering thermogenesis	Amount of thermogenin
Cold-adapted adult rat	Increases up to three weeks in the cold [5]	Increases up to three weeks in the cold [6]
Newborn guinea pig	Immediately maximum post-partum [5]	Maximum reached before 24 hours postpartum [53]
Newborn rat	Thermoneutral temperature decreases from day 5 postpartum [15]	Maximum at day 5 post-partum [60]
Newborn hamster	Poikilothermic until day 12 postpartum [55]	Large increase between days 12 and 20 post-partum [61]
Adult hamster, control, T_a 21°C	Relatively high capacity [69]	Relatively high concentration [62]
Diet-induced thermogenesis in rat	Elevated capacity [56]	Elevated [4]
Genetically obese mouse (ob/ob)	Impaired thermogenic response to cooling already at day 17 postpartum [68]	Decreased already at day 17 postpartum [25]
Diabetic obese mouse (db/db)	Reduced [25]	Reduced [25]
Hyperthyroid rat	Increased basal metabolic rate	Decreased [59]
Hypothyroid rat	Decreased basal metabolic rate	Increased *

*Sundin, personal communication.

the first days of adaptation there is a rapid increase in the concentration of thermogenin, but with time, the concentration starts to diminish, a condition also found by Desautels et al. [18]. If, however, one studies the capacity of rats to exhibit nonshivering thermogenesis (Fig. 6D), it is clear that this capacity reaches a plateau after three weeks and there is no sign of any decrease with time. Thus, at first glance, the correlation between thermogenin and thermogenesis seems poor.

During adaptation to cold, there is a successive increase in the total amount of mitochondria found in brown fat adipose tissue (Fig. 6B), as also found by Thomson et al. [67]. Thus the total amount of thermogenin, which is the product of the thermogenin concentration (nmol/mg mitochondrial protein) and the total amount of mitochondrial protein, remains at a constant level after the third

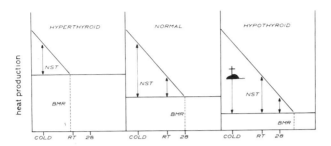

Fig. 7. Interaction between thyroid thermogenesis and brown fat thermogenesis. The figure summarizes the results from [59] and Sundin (personal communication). Note also the change in the lower critical temperature. RT = room temperature, BMR = basal metabolic rate, NST = nonshivering thermogenesis.

week (Fig. 6C). Two conclusions can be drawn: (1) that it is the total amount of thermogenin in the brown adipose tissue which determines the total thermogenic capacity of the tissue and (2) that there is a good correlation between the amount of thermogenin within an animal and the capacity of that animal for nonshivering thermogenesis.

This good correlation can be extended to a number of physiological and pharmacological states summarized in Table 2. It is also clear that to the physiological states in which brown fat is active which traditionally include the newborn mammal, the cold-adapted mammal and the arousing hibernator, should now be added "the well-fed mammal," i.e., one showing diet-induced thermogenesis.

The results of the interaction between thyroid thermogenesis and brown fat thermogenesis are best summarized as suggested in Fig. 7. Thus, under euthyroid conditions, i.e., "normal", there is much thermogenin (much brown fat thermogenesis) in the cold, and none at thermoneutral temperatures. In hyperthyroid rats the basal metabolism is increased (by mechanisms suggested by Edelman and Ismail-Beigi [21] to include stimulation of the cellular Na^+/K^+-ATPase) and the demand for thermogenesis at low temperatures is diminished and the amount of thermogenin therefore is lower [59]. In hypothyroid animals, e.g., methimazole treated, the basal metabolism is decreased; at a temperature, 28°C, which earlier represented thermoneutral conditions, the animals are now in demand of extra heat and there is more thermogenin (Sundin, personal communication).

Hypothyroid rats do not survive in the cold for prolonged periods. Whether this should be taken as an indication of a specific, permissive role of thyroid in the cold, or whether it is simply an effect of the combination of two different stresses is not known. It does seem essential for survival in cold that the animals are in good physical condition. For example, we have found that rats fed a diet deficient in vitamin B do not survive in the cold, although they are symptom-free at normal ambient temperatures.

The results of these pharmacological treatments of rats with thyroxine and methimazole have established the interaction between brown fat thermogenesis and thyroid thermogenesis at the two end points on the thyroid scale. These results can be interpolated to intermediate and probably more physiological alterations of thyroid activity. The relationships shown here are probably the most evident demonstrations of an "antibrown fat effect" of thyroid hormones. It must, however, be stressed that these effects of thyroid hormones on brown fat and especially on thermogenin are probably not direct but indirect effects, resulting from thyroid induced changes in basal heat production which secondarily regulates the activity of brown fat. The nature of this regulation is still unknown. It has been reported, for example, that chronic treatment with norepinephrine cannot increase the amount of thermogenin significantly [17].

Thermogenin and the role of acyl-CoA as a possible physiological effector of thermogenesis

Whereas the existence of thermogenin in the membrane of brown fat mitochondria endows the tissue with a mechanism for unlimited heat production, the thermogenin channel for protons cannot stay open in an unregulated way. This is obvious because, in contrast to thyroid thermogenesis ("basal metabolism"), brown fat thermogenesis (nonshivering thermogenesis) is a *facultative* heat production [29]. It is activated physiologically by cold and pharmacologically by norepinephrine injection. It is switched off physiologically at thermoneutral temperatures, and pharmacologically immediately at the end of the norepinephrine infusion. Thus, the activity of thermogenin, i.e., the openness of the channel for protons must be regulated. Metabolic inhibitors and activators must exist.

The *inhibitors* have been known for a long time: the purine nucleotides. ADP, ATP, GDP, and GTP can all bind to the binding site on thermogenin and, by doing this, close the proton channel, i.e., inhibit the activity of thermogenin [8, 50] (Fig. 8). The affinities of the nucleotides for the site is rather high, in the micromolar range. It has long been clear that the millimolar concentrations of these nucleotides outside the mitochondria, in the cytosol, are

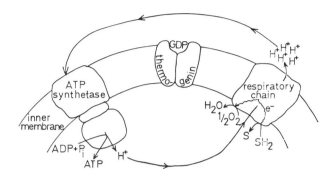

Fig. 8. Inhibition of the proton-current short circuit in brown fat mitochondria by purine nucleotides (GDP). When purine nucleotides are bound to thermogenin, the proton channel closes and proton reentry is limited by the capacity of other systems: the mitochondria are in an energy conserving state. From [9] by permission.

sufficiently high to close the proton channel of thermogenin [11]. Although it seems that brown fat mitochondria, through their thermogenin, have a potential capacity for unlimited respiration, in nature there would never be a situation where this capacity could be manifested.

In the search for a physiological activator for thermogenin and for thermogenesis, it was reasoned that the so-called "activated fatty acids," i.e., the acyl-CoAs, might be able to function as activators of thermogenin. The biochemical structure of acyl-CoA (Fig. 9) can be described as a purine nucleotide (ADP) linked by a somewhat complicated bridge to a fatty acyl group, palmitic acid. As a purine nucleotide derivative, this compound could be predicted to interact with the binding site on thermogenin, but the presence of the link and acyl-moiety makes the molecule quite bulky and thus may affect the conformation of thermogenin. Further, acyl-CoA would be produced at high rates when thermogenesis is stimulated.

The ability of acyl-CoA to act as a thermogenic effector in brown adipose tissue mitochondria has been tested. These experiments are summarized in Table 3.

As expected from their structure, acyl-CoAs could compete with GDP for the binding site on thermogenin. Of greater interest is the fact that they can reintroduce a high halide permeability in brown fat mitochondria, i.e., in principle they can activate thermogenin. It should be noted that this effect is

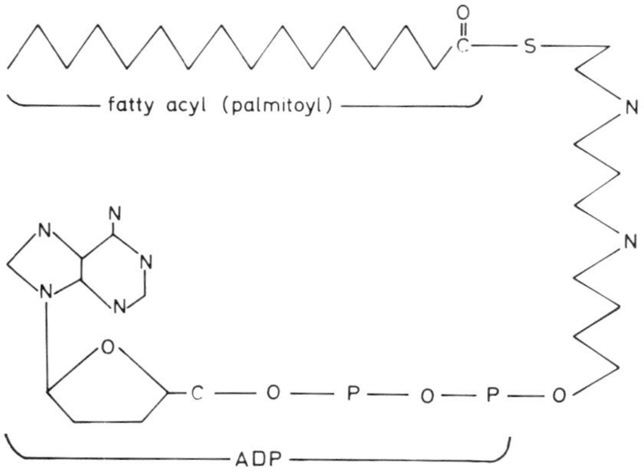

Fig. 9. The structure of acyl-CoA.

specific in two ways. First, only the permeability for certain anions (Cl$^-$ and Br$^-$) known to permeate through the thermogenin channel is increased, whereas the permeability of other anions, known to use other transport systems (phos-

TABLE 3

EFFECTS OF Acyl-CoA ON THE THERMOGENIN PROPERTIES OF BROWN FAT MITOCHONDRIA

| Thermogenin Properties | Effect of palmitoyl-CoA | Reversal by | | | Comments |
		purine nucleotides (GDP)	pyrimidine nucleotides (CDP)	chelators (EDTA)	
GDP binding	decrease [12]	yes [12]	not tested	always present [12]	
Halide (Cl-)	increase (in GDP-inhibited state) [12]	yes [12]	no [12]	no [12]	Also Br- permeable but to PO$_4^-$ [12] or K$^+$ [10]
Respiration (heat production) in coupled mitochondria	increase [10]	yes [10]	no [10]	no [10]	No similar effects on liver mitochondria [10]

phate), is unaltered. Second, the acyl-CoA activation can only be reversed by purine nucleotides, not by pyrimidine nucleotides or chelating agents, as would be expected from a specific interaction on the thermogenin site. In principle, the same specificity is seen if thermogenesis (oxygen consumption) is studied directly. These experiments are shown in Fig. 10.

In all three experiments, the mitochondria, for technical reasons, are brought into a coupled state by a combustion procedure and low amounts of a purine nucleotide are present [11, 31]. The mitochondria are then in a coupled state, i.e., after addition of a non-acyl-CoA substrate (acetyl-carnitine) no thermogenesis, i.e., increase in rate of oxygen consumption, occurs (Fig. 10a). Only if an artificial uncoupler, carbonyl cyanide p-trifluoromethoxyphenylhydrazone (FCCP), is added does respiration proceed unlimited and heat is evolved.

If, instead, acyl-CoA is added to the mitochondria, this substrate stimulates its own oxidation, i.e. it works as a thermogenic effector, and thermogenesis proceeds at a high rate (Fig. 10b, and Fig. 11). This is a specific effect of acyl-CoA on thermogenin because if a high amount of GDP is present acyl-CoA can no longer evoke its own oxidation (Fig. 10c).

Thus, in the question of specificity, acyl-CoA fulfills all criteria for being the thermogenic effector in brown adipose tissue.

This, however, is only as a demonstration that acyl-CoA could function as the effector, not that it is physiologically the thermogenic effector (the physiological uncoupler). At present, there is no experimental evidence for this, neither in isolated brown fat cells, nor in relation to the thermogenic response of the entire animal.

Clarification of the role of acyl-CoA in the regulation of thermogenesis is, as yet, an unsolved problem of the metabolism of brown fat. An even greater and highly challenging problem is the regulation of the synthesis of thermogenin and of the brown adipose tissue in parallel with the requirements of the animal for extra heat.

It seems that this regulation is not merely a question of a feedback mechanism which monitors the discrepancy between heat loss and heat production in the body and regulates the amount of thermogenin and total brown fat in accordance with current demands. Other mechanisms also must be present to account for diet-induced activation of brown fat and for preparation for winter (hibernation), which may be caused solely by day length changes [27], as well as to account for the ability to anticipate future thermogenic needs.

The clarification of these control mechanisms, their neural and hormonal background, and their metabolic effects on tissue, leaves us with a series of

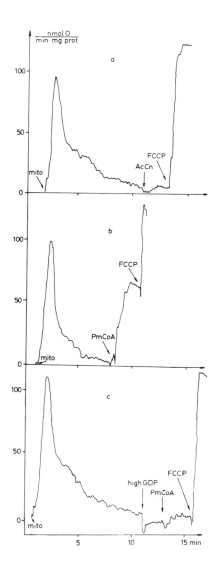

Fig. 10. Specific, partial uncoupling of brown fat mitochondria by palmitoyl-CoA.

a. AcCn = 5 mM acetyl-L-carnitine; FCCP = 0.2 µM carbonyl cyanide p-trifluoromethoxyphenyl - hydrazone.

b. PmCoA = 5 µM palmitoyl-CoA.

c. High GDP = 2 mM GDP.

Oxygen consumption was measured with a Clark-type Yellow Springs oxygen electrode probe 4004 in a final volume of 1 ml in a medium of 100 mM KCl, 20 mM K-TES, 4 mM KH_2PO_4, 2 mM $MgCl_2$, 1 mM EDTA, 3 mM malate, 1 mM ATP, 1 mM L-carnitine and 1.5 µM coenzyme A, pH 7.2 with 0.7 mg mitochondrial protein. The temperature was 23°C. From [11] by permission.

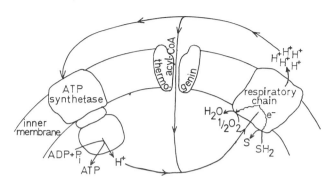

Fig. 11. Reintroduction of the proton-current short circuit in brown fat mito-chondria by acyl-CoA. When acyl-CoA is bound to thermogenin, it displaces purine nucleotides and reopens the proton channel. Respiration, and thus thermogenesis, can proceed unhindered by the building-up of a proton gradient. From [9] by permission.

problems of not only theoretical interest, but with possible practical implica-tions in areas as distant as energy conservation and control of obesity.

SUMMARY

When a hamster arouses from hibernation, it loses so much lipid from its brown fat depots that this nearly accounts for all heat produced. All lipid may not necessarily be combusted in the brown fat cells themselves, as cells both from control and cold-adapted hamsters are able to release fatty acids to the circulation. The cells from cold-adapted animals have, however, a lower norepi-nephrine sensitivity and thus may be in a metabolically desensitized state. The ability to produce heat in brown adipose tissue is dependent upon the presence in the mitochondria of a specific protein of molecular weight 32,000, for which we introduce the name *thermogenin*. The amount of *thermogenin* and the capacity for nonshivering thermogenesis are closely correlated in a series of different states, including hypo- and hyperthyroidism. The activity of thermogenin, i.e., thermogenesis, is inhibited by extramitochondrial purine nucleotides and the activity is increased by acyl-CoA which thus may be the physiological effector of thermogenesis in brown adipose tissue.

REFERENCES

1. Al-Shaikhaly, M. H. M., Nedergaard, J. and Cannon, B. (1979) Sodium-induced calcium release from mitochondria in brown adipose tissue. Proc. Natl. Acad. Sci. USA, 76, 2350-2353.

2. Bertin, R. (1976) Glycerokinase activity and lipolysis regulation in brown adipose tissue of cold acclimated rats. Biochimie, 58, 431-434.

3. Bertin, R. and Portet, R. (1976) Effects of lipolytic and antilipolytic drugs on metabolism of adenosine 3':5'-monophosphate in brown adipose tissue of cold acclimated rats. Eur. J. Biochem., 69, 177-183.

4. Brooks, S. L., Rothwell, N. J., Stock, M. J., Goodbody, A. E. and Trayhurn, P. (1980) Increased proton conductance pathway in brown adipose tissue mitochandria of rats exhibiting diet-induced thermogenesis. Nature, 286, 274-276.

5. Bruck, K. (1980) Nonshivering thermogenesis and brown adipose tissue in relation to age, and their integration in the thermoregulatory system, in Brown Adipose Tissue, Lindberg, O., ed. American Elsevier, New York, pp. 117-154.

6. Bukowiecki, L., Follea, N., Vallieres, J. and LeBlanc, J. (1978) β-adrenergic receptors in brown adipose tissue. Characterization and alterations during acclimation of rats to cold. Eur. J. Biochem., 92, 189-196.

7. Bukowiecki, L. and Lindberg, O. (1974) Control of sn-glycerol 3-phosphate oxidation in brown adipose tissue mitochondria by calcium and acyl-CoA. Biochim. Biophys. Acta, 348, 115-125.

8. Cannon, B. and Lindberg, O. (1979) Mitochondria from brown adipose tissue: Isolation and properties. Meth. Enzymol. 55F, 65-78.

9. Cannon, B. and Nedergaard, J. (1980) The function and properties of brown adipose tissue in the newborn, in Biochemical Development of the Fetus and the Neonate, Jones, C. T., ed. Elsevier, Amsterdam. (In press.)

10. Cannon, B., Nedergaard, J. and Sundin, U. (1980) Physiological uncoupling in brown fat mitochondria, in Proceedings of the International Symposium on Thermal Physiology, Pécs, Hungary. (In press.)

11. Cannon, B., Nicholls, D. G. and Lindberg, O. (1973) Purine nucleotides and fatty acids in energy coupling of mitochondria from brown adipose tissue, in Mechanisms in Bioenergetics, Azzone, G. F., et al., eds. Academic Press, New York and London, pp. 357-364.

12. Cannon, B., Sundin, U. and Romert, L. (1977) Palmitoyl coenzyme A: A possible physiological regulator of nucleotide binding to brown adipose tissue mitochondria. FEBS Lett., 74, 43-46.

13. Chinet, A., Clausen, T. and Girardier, L. (1977) Microcalorimetric determination of energy expenditure due to active sodium-potassium transport in the soleus muscle and brown adipose tissue of the rat. J. Physiol. (Lond.), 265, 43-61.

14. Chinet, A. and Durand, J. (1979) Control of the brown fat respiratory response to noradrenaline by catechol-O-methyltransferase. Biochem. Pharmacol., 28, 1353-1361.

15. Conklin, P. and Heggeness, F. W. (1971) Maturation of temperature homeostasis in the rat. Am. J. Physiol., 220, 333-336.

16. Depocas, F. (1960) The calorigenic response of cold-acclimated white rats to infused noradrenaline. Can. J. Biochem. Physiol., 38, 107-114.

17. Desautels, M. and Himms-Hagen, J. (1979) Roles of noradrenaline and protein synthesis in the cold-induced increase in purine nucleotide binding by rat brown adipose tissue mitochondria. Can. J. Biochem., 57, 968-976.

18. Desautels, M., Zaror-Behrens, B. and Himms-Hagen, J. (1978) Increased purine nucleotide binding, altered polypeptide composition and thermogenesis in brown adipose tissue mitochondria of cold-acclimated rats. Can. J. Biochem., 56, 378-383.

19. Dorigo, R., Gaion, R. M. and Fassina, G. (1974) Lack of correlation between cyclic AMP synthesis and free fatty acid release in brown fat of cold-adapted rats. Biochem. Pharmacol., 23, 2877-2885.

20. Durand, J., Giacobino, J. -P., Deshusses, J. and Girardier, L. (1979) Re-evaluation of the relationship between catechol-O-methyl transferase and the binding of norepinephrine to brown adipocyte membranes. Biochem. Pharmacol., 28, 1347-1351.

21. Edelman, I. S. and Ismail-Beigi, F. (1974) Thyroid thermogenesis and active sodium transport. Recent Prog. Hormone Res., 30, 235-257.

22. Feist, D. D. and Quay, W. B. (1969) Effects of cold acclimation and arousal from hibernation on brown fat lipid and protein in the golden hamster (Mesocricetus auratus). Comp. Biochem. Physiol., 31, 111-119.

23. Fiak, S. A. and Williams, J. A. (1976) Adrenergic receptors mediating depolarization in brown adipose tissue. Am. J. Physiol., 231, 700-706.

24. Girardier, L., Seydoux, J. and Clausen, T. (1968) Membrane potential of brown adipose tissue. J. Gen. Physiol., 52, 925-940.

25. Goodbody, A. E. and Trayhurn, P. (1980) Functional changes in brown adipose tissue of the diabetic-obese (db/db) mouse. International Symposium on Thermal Physiology. Pécs, Hungary. (In press.)

26. Hayward, J. S. and Lyman, C. P. (1967) Nonshivering heat production during arousal from hibernation and evidence for the contribution of brown fat, in Mammalian Hibernation III, Fisher, K. C. , et al., eds. American Elsevier, New York, pp. 325-355.

27. Heldmaier, G. and Steinlechner, S. (1981) Seasonal control of thermogenesis by photoperiod and ambient temperature in the djungarian hamster, Phodopus songurus. Crybiology, 18, 96-97.

28. Himms-Hagen, J. (1972) Lipid metabolism during cold-exposure and during cold-acclimation. Lipids, 7, 310-323.

29. Himms-Hagen, J. (1978) Biochemical aspects of nonshivering thermogenesis, in Strategies in Cold: Natural Torpidity and Thermogenesis, Wang, L. C. H. and Hudson, J. W., eds. Academic Press, New York and London, pp. 595-617.

30. Hittelman, K. J., Bertin, R. and Butcher, R. W. (1974) Cyclic AMP metabolism in brown adipocytes of hamsters exposed to different temperatures. Biochim. Biophys. Acta, 338, 398-407.

31. Hittelman, K. J., Lindberg, O. and Cannon, B. (1969) Oxidative phosphorylation and compartmentation of fatty acid metabolism in brown fat mitochondria. Eur. J. Biochem., 11, 183-192.

32. Horowitz, J. M., Horwitz, B. A. and Smith, R. Em. (1971) Effect in vivo of norepinephrine on the membrane resistance of brown fat cells. Experientia, 27, 1419-1421.

33. Horwitz, B. A. and Eaton, M. (1975) The effect of adrenergic agonists and cyclic AMP on the Na$^+$/K$^+$ ATPase activity of brown adipose tissue. Eur. J. Pharmacol., 34, 241-245.

34. Joel, C. D. (1965) The physiological role of brown adipose tissue, in Handbook of Physiology, sect. 5, Adipose Tissue, Reynold, A. E. and Cahill, G. F., Jr. eds. Am. Physiol. Soc., Washington, D.C., pp. 59-85.

35. Kramar, R., Huttinger, M., Gmeiner, B. and Goldenberg, H. (1978) β-oxidation in peroxisomes of brown adipose tissue. Biochim. Biophys. Acta, 531, 353-356.

36. Lin, C. S., Hackenberg, H. and Klingenberg, M. (1980) The uncoupling protein from brown adipose tissue mitochondria is a dimer. A hydrodynamic study. FEBS Lett., 113, 304-306.

37. Lin, C. S. and Klingenberg, M. (1980) Isolation of the uncoupling protein from brown adipose tissue mitochondria. FEBS Lett., 113, 299-303.

38. Mohell, N., Nedergaard, J. and Cannon, B. (1980) An attempt to differentiate between α- and β-adrenergic respiratory responses in hamster brown fat cells, in International Symposium on Thermal Physiology, Pécs, Hungary. (In press.)

39. Nedergaard, J. (1980) Control of fatty acid utilization in Brown Adipose Tissue. University of Stockholm, ISBN 91-7146-089-6.

40. Nedergaard, J. (1981) Effects of cations on brown adipose tissue in relation to possible metabolic consequences of membrane depolarisation. Eur. J. Biochem. (In press.)

41. Nedergaard, J. (1981) Recovery of catecholamine sensitivity in brown fat cells isolated from cold-adapted hamsters and rats. Am. J. Physiol. (In Press.)

42. Nedergaard, J., Alexson, S. and Cannon, B. (1980) Cold adaptation in the rat: Increased brown fat peroxisomal β-oxidation relative to maximal mitochondrial oxidative capacity. Am. J. Physiol. (Cell Physiol.), 239. C208-216.

43. Nedergaard, J., Al-Shaikhaly, M. H. M. and Cannon, B. (1979) Sodium-induced calcium release from mitochondria in brown fat, in Function and Molecular Aspects of Biomembrane Transport, Quagiliariello, E. et al., eds., Elsevier/ North Holland, pp. 175-178.

44. Nedergaard, J. and Cannon, B. (1980) Effects of monovalent cations on Ca^{2+} transport in mitochondria; a comparison between brown fat and liver mitochondria from rat. Acta Chem. Scand., B34, 149-151.

45. Nedergaard, J. and Cannon, B. (1980) A possible metabolic effect of membrane depolarisation in brown adipose tissue, in International Symposium on Thermal Physiology, Pécs, Hungary. (In press.)

46. Nedergaard, J., Cannon, B. and Lindberg, O. (1977) Microcalorimetry of isolated mammalian cells. Nature, 267, 518-520.

47. Nedergaard, J. and Lindberg, O. (1979) Norepinephrine-stimulated fatty-acid release and oxygen consumption in isolated brown fat cells. Eur. J. Biochem, 95, 139-145.

48. Nedergaard, J. and Lindberg, O. (1981) The brown fat cell. Int. Rev. Cytology. (In press.)

49. Nedergaard, J., Nanberg, E. and Cannon, B. (1981) Cationic regulation of thermogenesis in brown adipose tissue, in Proceedings of International Sym-

posium Survival in Cold, Jansky, L. and Musacchia, X. J., eds. Acta Universitatis Carolinae, Prague. (In Press.)

50. Nicholls, D. G. (1979) Brown adipose tissue mitochondria. Biochim. Biophys. Acta, 549, 1-29.

51. Pettersson, B. and Vallin, I. (1976) Norepinephrine-induced shift in levels of adenosine 3':5'-monophosphate and ATP parallel to increased respiratory rate and lipolysis in isolated hamster brown-fat cells. Eur. J. Biochem., 62, 383-390.

52. Rabi, T., Cassuto, Y. and Gutman, A. (1977) Lipolysis in brown adipose tissue of cold- and heat-acclimated hamsters. J. Appl. Physiol., 43, 1007-1011.

53. Rafael, J. and Heldt, H. W. (1976) Binding of guanine nucleotides to the outer surface of the inner membrane of guinea pig brown fat mitochondria in correlation with the thermogenic activity of the tissue. FEBS Lett., 63, 304-308.

54. Ricquier, D., Gervais, C., Kader, J. C. and Hemon, Ph. (1979) Partial purification by guanosine-5'-diphosphate-agarose affinity chromatography of the 32,000 molecular weight polypeptide from mitochondria of brown adipose tissue. FEBS Lett., 101, 35-38.

55. Rink, R. D. (1969) Oxygen consumption, body temperature and brown adipose tissue in the postnatal golden hamster (Mesocricetus auratus). J. Exp. Zool., 170, 117-123.

56. Rothwell, N. J. and Stock, M. J. (1979) A role for brown adipose tissue in diet-induced thermogenesis. Nature, 281, 31-35.

57. Skala, J. P. and Knight, B. L. (1977) Protein kinases in brown adipose tissue of developing rats. J. Biol. Chem., 252, 1064-1070.

58. Skala, J. P. and Knight, B. L. (1979) Cyclic GMP and cyclic GMP-dependent protein kinase in brown adipose tissue of developing rats. Biochim. Biophys. Acta, 582, 122-131.

59. Sundin, U. (1981) GDP-binding to rat brown fat mitochondria. The influence of thyroxine under different ambient temperatures. Am. J. Physiol. (In press.)

60. Sundin, U. and Cannon, B. (1980) GDP-binding to the brown fat mitochondria of developing and cold-adapted rats. Comp. Biochem. Physiol., 65B, 463-471.

61. Sundin, U., Herron, D. and Cannon, B. (1981) Brown fat thermoregulation in developing hamsters (Mesocricetus auratus): A GDP-binding study. Biol. Neonate. (In press.)

62. Sundin, U., Moore, G. and Cannon, B. (1981) GDP-binding and cytochrome concentration in brown fat mitochondria from hamsters adapted to different temperatures. Cryobiology, 18, 105.

63. Svartengren, J., Mohell, N. and Cannon, B. (1980) Characteristics of (^{3}H)-dihydroergocryptine binding sites on intact brown adipocytes. International Symposium on Thermal Physiology, Pécs., Hungary. (In press.)

64. Svartengren, J., Mohell, N. and Cannon, B. (1980) Characterization of (^{3}H)-dihydroergocryptine binding sites in brown adipose tissue. Evidence for the presence of α-receptors. Acta Chem. Scand., B34, 231-232.

65. Svartengren, J., Svoboda, P. and Cannon, B. (1981) β-adrenergic receptors in brown fat. A comparison between cold-adapted and control hamsters, in

Proceedings of International Symposium Survival in Cold, Jansky, L. and Musacchia, X. J., eds. Acta Universitatis Carolinae, Prague. (In Press.)

66. Svoboda, P., Svartengren, J., Snochowski, M., Houstěk, J. and Cannon, B. (1979) High number of high-affinity binding sites for (-)-(^{3}H)dihydroalprenolol on isolated hamster brown-fat cells. A study of the β -adrenergic receptors. Eur. J. Biochem., 102, 203-210.

67. Thomson, J. F., Habeck, D. A., Nance, S. L. and Beetham, K. L. (1969) Ultrastructural and biochemical changes in brown fat in cold-exposed rats. J. Cell Biol., 41, 312-334.

68. Trayhurn, P., Thurlby, P. L. and James, W. P. T. (1977) Thermogenic defect in pre-obese ob/ob mice. Nature, 266, 60-62.

69. Vybiral, S. and Jansky, L. (1972) Thermoregulatory significance of catecholamine thermogenesis in golden hamsters. Physiol. Bohemoslov., 21, 121-122.

NEUROENDOCRINE ASPECTS OF HIBERNATION

HENRY SWAN
Department of Clinical Sciences, Colorado State University, Ft. Collins,
Colorado 80521, U.S.A.

INTRODUCTION

This review analyzes some of the more specific relationships of neuroendocrine agents to the induction of hibernation without considering the
associated general polyglandular endocrine changes. Endocrinology in hibernation has been frequently reviewed [26, 30, 32, 40] and general considerations
will not be stressed.

Hibernation is regarded as an active physiologically controlled state
characterized by suppression of metabolic rate, altered control of thermoregulating mechanisms, a drop in body temperature to just above ambient temperature,
torpor, the capability of body tissues to remain functional at whatever hypothermic temperatures result, and, finally, the ability of thermogenesis to
rewarm the body upon arousal [22, 31, 62, 70]. In terms of metabolism, it is
universally agreed that oxygen consumption $\dot{V}_{O_2}$ is depressed before body temperature falls [47, 48, 49, 65].

Even before this information was available, suggestions were made that some
controlling hormone might be responsible for the induction of hibernation. In
1932, Nitschke and Maier [55] injected an extract of a hibernator's lymphatic
tissue into nonhibernators and hibernators, and produced a physiological state
in which body temperature (T_b), oxygen consumption ($\dot{V}_{O_2}$) and blood sugar were
reduced. Wendt [71], in 1937, apparently was the first to use extracts of brown
adipose tissue (BAT). Extracts from hibernating hedgehogs, injected intra-
peritoneally, lowered the metabolic rate of rats. Hook [27] made similar
observations in 1940. In 1952, Kroll [44] pinpointed the idea by asking, "Is
there a sleep substance in hibernation?" The following year he described experiments in which a sleep-like state was observed in cats following injection of
brain extracts from hibernating hamsters and hedgehogs [45]. The resemblance to
hibernation was slight since body temperatures did not fall significantly and
additive extracts of BAT were without effect.

Three years later, Zirm [73] reported that subcutaneous transplantation or
injections of extracts of BAT from hibernating hedgehogs into mice produced a

lowering of T_b and a fall in metabolism. No such effect was seen with BAT extracts from nonhibernating hedgehogs or from tissues other than BAT.

These studies, apparently supported Wendt's observations of 20 years earlier, strongly advanced the idea that BAT was primarily glandular in function and was the source of the "hibernating hormone" [37]. Many laboratories began a vigorous search for "hybrin" or "hibernin" as it was variously called. Morrison and Allen [53], Bigelow et al. [5], and Johannsen [38] were unable to confirm any effect of injections of brown fat extracts from hibernating animals in non-hibernators. Meanwhile, the extraordinary heat producing capacity of BAT in newborn rabbits and human infants and its increased thermogenic capabilities as an adaptation to cold was receiving close attention [3, 13, 14, 53]. Frequent reviews of BAT as a thermogenic organ appeared [46, 60, 61]. In general, investigators agree that BAT is not the source of a hormone which induces hibernation, but that it is an important thermoregulating tissue both in hibernating and in nonhibernating animals.

Neurohumoral activity of the hypothalamus, the well-documented source of norepinephrine (NE) and serotonin (5-HT), was being investigated, not only in sleep [39, 43] but also in hibernation. In 1963, I postulated the existence of an antimetabolic hormone [63] which was later termed "antabolone." In 1969, Swan et al. [67], presented the first evidence that an extract from the brain of aestivating lungfish, *Protopterus aethiopicus,* when injected intravenously into white rats, induced a state of torpor associated with a fall in $\dot{V}_{O_2}$ of 35%, followed by a drop in body temperature of 3°C. These effects were not seen following injection of extracts from the brains of active fish. Subsequently, additional evidence of the existence of antabolone was presented, this time using extracts from the brains of hibernating ground squirrels, *Citellus tridecemlineatus* [68]. Again, intravenous injection into white rats was followed by a decrease in $\dot{V}_{O_2}$ of about 30% and a fall in body temperature of 3°C. Extracts from active, normothermic squirrel brains caused minimal responses in the same direction.

From 1976 to the present, Reinhard (personal communication) in Munich has been repeating lungfish studies using large numbers of aestivating and non-aestivating lungfish brains harvested and extracted in Chad, Central Africa. Using the body temperature response in young mice as the bioassay technique, he confirms that hypothermia of 2°-4°C follows intravenous injection of extracts from aestivating fish, but only slight effect was seen following extracts from active fish. Partially purified, the active agent was a peptide between 500 and 1500 daltons destroyed by Pronase-P.

In any event, the existence of an agent, extractable from the brai
animals, which is capable of suppressing metabolic rate in the ι
sleeping state, has now been documented by more than a single group of investigators. The question may be asked, "What level of metabolism can be suppressed safely without crippling the energy transduction necessary for life?"

The nature of obligate thermogenesis

At best, 25% of the oxygen consumed in oxidative phosphorylation results in
high energy phosphate bonds ($\sim P$). The remaining energy is converted to heat, a
form of energy not reconvertable to work. The control of this system, which
continuously releases heat, is geared to the need of the tissue for ATP to
perform work; it is not governed by nor is it responsive to T_a or T_b.

This is the important point which is commonly forgotten. There is self-
regulated oxidative metabolism going on in the tissues at all times which
converts energy to a form useable to accomplish the essential work of the
resting cells and thus keeps the tissues alive. To further develop this line of
discussion, I use the term M_{EE} to mean "essential energesis." Considered as a
machine to do work, it defines the minimum viable metabolic rate at various
temperatures. It is mass dependent, and can not be turned off. It contributes
heat within parameters of the basal metabolic rate (BMR), but it contributes
heat as a byproduct. Coupled oxidative phosphorylation is not a process controlled primarily to produce heat; in this sense it is not "thermogenesis,"
though it is thermogenic.

The byproduct heat of M_{EE} is not adequate to maintain homeothermy, particularly in small animals. Recognition of this is essential to formulating a
useable concept of the heat components of BMR and their relationship to nonshivering thermogenesis (NST). In 1938, Keller [41], using differential
coagulation, produced hypothalamic lesions in dogs. Some of the dogs became
poikilothermic, and lost their ability to thermoregulate; others, in addition,
exhibited a marked fall in BMR. Keller [42] remarked "analysis of these differential tissue defects and associated deficits forces the conclusion that
physiological regulation against cold is primarily a function of the 'neural
hypothalamus,' whereas the normally elevated basal energy metabolism is an
endocrine affair, presumably a function of the 'endocrine hypothalamus'." His
experimental animals lost 40% of their BMR due to the hypothalamic lesions and
had become poikilothermic. Their body temperatures declined with the slightest
cold exposure. But they were alive, and if protected from cold, they survived.
He stated "analysis of the foregoing data should leave no doubt as to the

existence of a nonshivering heat source in the dog." This may be taken to be the first important evidence that BMR contains a proportion of NST regulated by the hypothalamus.

Since that time much study of the nature and control of NST has been done, but agreement on the meaning of the term is still lacking [34]. Hsieh et al. [29] suggested that NST could be divided into two components, that which is obligatory, and that which is regulatory.

The biochemical mechanism of NST in BAT was reviewed in depth during earlier symposia in Prague [33, 36]. There was much discussion on the mechanism of non-shivering thermogenesis based on loosening of coupling or alternate uncoupling involving such variables as: triglyceride hydrolysis; free fatty acids as sub-strates and as uncoupling agents; activation of fatty acids; Na^+ and K^+ membrane flux; norepinephrine (NE) control mechanisms and dose responses; the relationship of NST to thyroidectomy; and the separation between spinal cord control of shivering from hypothalamic control of NST. A concept of NST as a specific definable heat generating mechanism did not emerge.

By 1973, Jansky [34] in his review recognized and defined nonshivering thermogenesis "as a specific heat producing mechanism due to processes which do not involve muscular contractions. Thus the heat production under the con-ditions of basal metabolism is mostly NST, and is called obligatory or basal NST." However, his survey was devoted entirely to consideration of regulatory NST. There were no data on obligatory NST to review.

During the next four years, sufficient progress had been made in under-standing the biochemical mechanisms of NST that a general consensus seemed to have been achieved. The reviews by Cannon, et al. [9], Himms-Hagen [24] and Horwitz [28] were complementary and exhaustive. They agreed that facultative NST is achieved in rat, guinea pig and hamster in brown fat and skeletal muscle. The biochemical reactions occur primarily in mitochondria, and to some extent in the cytoplasm, characterized either by loose coupling, as a result of lipolysis and free fatty acids, or by increased ATPase membrane translocation of ions (a process called ion cycling). Both mechanisms are induced by endogenous or exogenous NE via cAMP. These three reviews were confined to discussions of regulatory NST. The data presented were derived either from cold adaptation or injection of NE in vitro or in vivo. No data were presented concerning obligatory NST or its control mechanisms.

More recently, the role of loosely coupled mitochondrial proton conductance pathway sites has received increased attention [9, 15, 25, 54,] as a facultative thermogenic mechanism. Purine nucleotide binding on the mitochondrial mem-

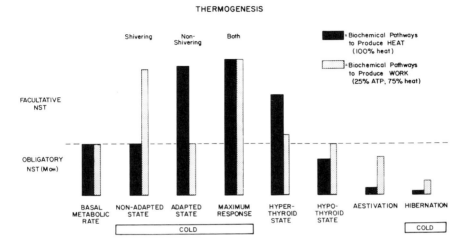

Fig. 1. The heat contribution of nonshivering thermogenesis in various physiologic states including hibernation.

brane's 32,000 polypeptide sites is rapidly increased by either cold exposure or
infusion of NE in the rat.   A concomitant slower response, also mediated by NE,
is tissue hypertrophy requiring cytosolic but not mitochondrial protein
synthesis.   The changes in mitochondrial composition which unmask the proton
conductance sites, however, are not mediated by NE. It is possible that further
understanding of this control mechanism might disclose the cybernetics of
obligatory NST.

Himms-Hagen [24] said "The control mechanisms for facultative thermogenesis
would be expected to differ from the control mechanism for obligatory thermogenesis, although both would of course ultimately depend upon the same biochemical machinery."   There should be general accord with Himms-Hagen, i.e.,
that NST is indeed a specific biochemical machine to produce heat and should be
called NST whether it is part of BMR or superimposed upon it.   Even during
sleep, it constitutes a vital component of BMR.   Its control mechanisms are
poorly understood and should be an area of intense study, particularly in
hibernating animals.   NST is illustrated in various physiologic states in Fig.
1.   There is ample reason to suggest that control mechanisms require future
investigation.   The recent study of Grubb and Folk [21] suggesting that the NST
of BMR is regulated by alpha-adrenergic pathways may be the opening wedge in
this area of inquiry.

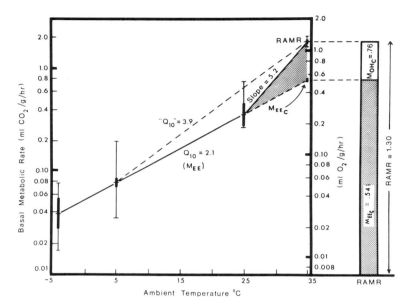

Fig. 2. Analysis of semi-log plot from Henshaw [23] on metabolic rates of bats
at various temperatures entering and during hibernation.   When the plot is a
straight line from 5°C and 25°C, it represents basic energy metabolism (M_{EE}) and
reflects the effects of temperature on $\dot{V}_{O_2}$ (Q_{10} of 2.1). From 25°C to 35°C, NST
is stimulated and the slope increases to 5.2.   Projection of the M_{EE} line to
35°C (coenothermia) shows that of the total BMR of 1.3 ml O_2/g/hr, 0.54 ml is
M_{EE} and 0.76 is M_{OH} (adapted from Swan [65]).

The components of basal (sleeping) metabolic rate

A physiological analysis of the thermogenic components of BMR and their
determinants was made a few years ago [64, 67].   This solution was derived by
analysis of data on the respiration of hibernating bats at various body temperatures as reported by Henshaw [23].   Between the temperatures of 5°C and 25°C
(Fig. 2) the animals are heterothermic and metabolism varies with temperature
along a Q_{10} line with a slope of 2.1.   This represents the weight-dependent
metabolism of essential energy consumption (M_{EE}).   If one extends this line
diagrammatically to 35°C, the temperature of coenothermia, it will define the
amount of metabolism necessary for cellular work at that temperature.   It is
weight dependent with a power function of 1.

Between 25°C and 35°C, however, the measured metabolism increased at a
different and much greater rate than M_{EE}.   In Fig. 2, the bar at the right is
the measured thermoneutral BMR. The difference between BMR and M_{EE} is a measure

of the metabolism necessary to maintain T_b over and above the heat produced as a byproduct of M_{EE}. This extra obligatory metabolism is surface dependent with a power function of 0.76 and is referred to as the metabolism of obligate heat (M_{OH}).

Thus the resting metabolism of these bats is composed of about 40% weight-dependent coupled metabolism and 60% surface-dependent NST. At rest, particularly when sleeping, all homeotherms are obligate heat producers:

$$RAMR = M_{EE} + M_{OH}.$$

The proportion of the resting metabolism which is NST (M_{OH}) varies greatly with size. The smaller the animal, the greater is its heat losing surface. A graph of the two components of RAMR of hibernators in a log-log plot would show that at four grams, all resting metabolism would be obligate heat. Thus the smallest hibernator would have to weigh more than four grams. *P. pepistrellus* at 7.5 grams is in fact the smallest hibernator known. At about 8 kg, the two components are equal. M_{OH} starts falling rapidly at 40 kg and at about 110 kg it will essentially disappear. Larger than this, one could speculate, a hibernator would die of hyperthermia from the heat of M_{EE} unless its shape were that of a rug.

These considerations contribute to an understanding of why only small animals are able to hibernate. If obligatory NST is indeed what is depressed, then there must be enough of it to allow a useful saving in the thermodynamic balance sheet. It would appear that hibernation will probably be limited to a weight range of 8 to 2,500 grams (hazel mouse to marmot). Within this range, M_{OH} comprises from 70-95% of resting metabolism.

This computation compares rather dramatically with the measured energy savings in the Richardson's ground squirrels reported by Wang [70]. Savings of 82-96% occurred during August to February, with an economy of 88% for the torpor season as a whole. In a ground squirrel, weighing 150 gm, M_{OH} is 85% of RAMR. Thus this compositional analysis of RAMR has received strong experimental support from Wang's 1978 study. In summary, the physiological determinants of obligatory NST indicate that it comprises 70-95% of the resting metabolism of hibernators. It is suggested that this is the component of BMR which is suppressed during entry into hibernation. Studies of the biochemical control mechanisms involved are prime areas for future research.

Serotonin

Although the neurotransmitter NE is evidently involved in arousal from hibernation, one may ask what is currently known about other bioamines of the hypothalamus during hibernation? In particular, 5-hydroxytryptamine (5-HT) appears to be most promising.

It is well established [35] that 5-HT turnover in the brain is increased in hibernation. Feldberg and Myers [16] demonstrated that injection of NE, E, and 5-HT into the cerebral ventricles and/or intrahypothalamically elicited significant changes in body temperature in the cat. They hypothesized that the hypothalamic balance between NE and 5-HT was a basic mechanism of thermoregulation. The importance of these bioamines in thermoregulation has been further corroborated in the rat [17].

The subject was reviewed by Jansky [35] and it was reported that, although optimum levels of 5-HT in the brain are necessary for the onset and maintenance of hibernation, nonetheless, central injection of 5-HT into alert ground squirrels did not significantly reduce T_b nor did it induce the onset of hibernation no matter what the season. Indeed, Beckman and Satinoff [4] caused premature arousal by hypothalamic injection of both NE and 5-HT. Glass and Wang [19] presented evidence that 5-HT influences thermoregulatory events only during arousal.

On the other hand, in nonhibernating species there is considerable evidence that the release of gonadotrophic and thyroid stimulating hormone is suppressed by increased activity of serotonergic pathways. It is tempting to suggest that 5-HT may in some fashion set the stage for entry into hibernation by modulating the overall endocrine pattern. Such a theory lacks documentation and is entirely speculative.

Trigger

The idea that a humoral agent may alter the general set of the neuroendocrine system, and thus facilitate the onset of hibernation, was given strong stimulus by Dawe and Spurrier [12]. In 1977 Dawe [11] presented a theory to explain their findings. They reported that dialysates of blood or plasma from hibernating *C. tridecemlineatus* or *M. monax* preserved in cold for up to six months, when injected i.v., induced summer hibernation in *C. tridecemlineatus*. Hibernation, so induced, consists of bouts which continue sporadically until the animal dies. Final arousal apparently rarely occurs.

Attempts to duplicate the induction of summer hibernation in various other species has been unsuccessful [1, 18, 52, 59]. In 1977 Dawe attributed the lack of success by other investigators to their failure to observe critical details in the serum collection technique, to be sure that cold adaptation had not occurred in the recipients, and to control certain environmental factors for both donor and recipients.

Oeltgen et al. [57] have explored a variety of techniques attempting to concentrate, purify, and ultimately to identify the chemical structure of the trigger. He and his coworkers believe that trigger is a small, thermolabile protease-sensitive protein in excess of 5,000 daltons. Apparently, it is closely associated with the albumen fraction of blood. The subject is reviewed and presented as a chapter in this volume.

Bioassay has been a continuing problem. Indirect methods such as the seasonal characterization of hemoglobin [56] have been of limited value. To date trigger has not shown an immediate or direct effect upon oxygen consumption, body temperature, or heart rate, but rather exerts its delayed effects in *C. tridecemlineatus* only through induced mechanisms. This limits the whole animal bioassay to the cumbersome and precisely programmed protocols of summer induced hibernation.

Recently, Meeker et al. [51], suggested bioassay by central injection of woodchuck trigger into monkeys, causing a recognizable response characterized by suppression of feeding, slowing of heart rate, and some drop in T_b. The assay, however, is not specific for trigger if it observes immediate depression in $\dot{V}_{O_2}$ and T_b. Such responses are more characteristic of bombesin, neurotensin, and other substances. Thus, for trigger, despite the use of such sophisticated laboratory techniques as isoelectric focusing, isotachophoresis, and affinity chromatography, no completely homogenous preparation suitable for molecular weight estimations, amino acid analysis, or sequence studies has been reported.

Antimetabolic neurohormones

Neurotensin: Neurotensin, a 13-amino-acid peptide initially obtained from the bovine hypothalamus [10], also identified in rat, mouse, rabbit, and human extracts, is reported to have a wide distribution in the CNS. Intravenous injection in rats can cause hypotension and hyperglycemia, effects which are not altered by adrenalectomy, hypophysectomy, or α-and β-adrenergic blockers. Since its effects on guinea pig ileum and rat duodenum are similar to bradykinin, neurotensin is classified as a kinin. When given intracisternally (i.c.) but not intravenously to mice at 25°C, neurotensin caused a fall in T_b of 3°C, reaching a maximum at 30 minutes. In a 4°C environment, the effect was much more profound and not maximal until 90 minutes. Equivalent i.c. doses of other hypothalamic peptides, namely, thyrotropin releasing factor (TRF), luteinizing releasing factor (LRF), the oxytocin derivative (MIF) and the peptide neurotransmitter substance P (SP) failed to cause a significant fall in T_b [6, 8, 69]. These studies did not include measurements of $\dot{V}_{O_2}$. Although neurotensin causes hypothermia, hypometabolism has not been documented.

Bombesin: Bombesin, a 14-amino-acid peptide originally isolated by Anastasi et al. [2] in 1972 from frog skin, like neurotensin, lowers blood pressure, produces gut contraction, and elevates plasma levels of glucagon, glucose, growth hormone, and prolactin. It is also distributed in mammalian CNS. Brown et al. [8] found that bombesin in a dose of 0.1 μg injected i.c. in 4°C cold exposed rats was ten thousand times more potent than neurotensin in producing hypothermia. Again, the peptide produced hypothermia only when injected intracisternally.

Of great interest was the subsequent study by the same authors [7] that TRF, prostaglandin E_2 (PGE$_2$) and naloxone reversed the hypothermic effect of bombesin. Similarly, the hyperthermia produced in rats at room temperature by TRF and PGE$_2$ was prevented by bombesin. Potent interrelated central effects of these substances on thermoregulation are thus demonstrated. Moreover, an opiate dependent step for the mechanism of action by bombesin may exist [58].

Unfortunately, no oxygen consumption studies were made during any of these studies so the relationship of changes in T_b to $\dot{V}_{O_2}$ was not demonstrated. However, a recently completed study by Wunder et al. [72] demonstrated this effect. Injections of 1 μg into the lateral ventricle or of 0.1 μg into the anterior hypothalamus (POAH) caused significant decrease in $\dot{V}_{O_2}$ and T_b. When injected into posterior hypothalamus, no such effects are seen. Of particular interest to us were the time relationships between the depression of $\dot{V}_{O_2}$ and that of T_b. In the rat, following injection of 0.1 μg bombesin into the POAH, $\dot{V}_{O_2}$ reaches maximum depression at 20 minutes just as T_b begins to fall. Maximum depression in T_b does not occur until an hour later.

We believe these data indicate that a primary suppression of metabolic rate comprises at least one of the mechanisms for the depression in body temperature caused by bombesin.

Endogenous opiates: The suggestion that bombesin may have an opiate-related step because of its reversal by naloxone may represent the only evidence for a relationship between the endorphins and thermoregulation. Nalaxone had no effect on temperature control in cold stressed rats [20]. Margulis et al. [50] observed an increase in heart rate in hibernating Turkish hamsters with 0.5 mg/kg of naloxone subcutaneously, but arousal was not precipitated unless the animals were approaching the end of a normal bout.

In 1977 a functional role for nalaxone was investigated in our laboratory. Nine hibernating ground squirrels (*C. tridecemlineatus*) were injected i.p. using a 27-guage needle; six animals were given 0.2 ml of cold saline and three animals were given cold saline with 0.08 mg. of naloxone. Every effort was made not to arouse the animals by handling or physical disturbance. Of the nine animals, however, three of the saline treated animals aroused from hibernation. None of the three nalaxone treated animals aroused.

There is therefore little evidence to date that the endorphins are significantly involved in either hibernation or thermogenesis.

Antabolone: As a result of early experience with fractionation by molecular weight we have consistently used the fraction which is greater than 1000 but less than 10,000 daltons as obtained using Amicon filtration.

The isolated perfused rat heart has been used to study whether the non-hibernating mammalian heart might be a target organ for the antimetabolic effects of hibernating *C. tricedemlineatus* brain extracts. A series of perfusions of the isolated beating rat heart were performed [66]. Brain extracts from active *C. tridecemlineatus*, as well as from antelope ground squirrel, guinea pig, mouse, and rat, were included in the study [66]. Hibernating ground squirrel brain extract increased the rate of coronary perfusion. However, the same effect was seen for all the other heterologous extracts. Homologous extracts from rat brain had an opposite but not significant effect. Associated with the increase in coronary flow was a marked increase in $\dot{V}_{O_2}$. Again, the rat brain extract produced an opposite effect. In concomitant studies, the effect on heart rate was not significant, the perfusion pressure remained constant, and the measured tension of the contractions was not significantly changed. In other words, there was no evidence of increased work performed by the heart.

We concluded that the rat heart was not a target organ for the antimetabolic activity of hibernating ground squirrel brain extracts. However, we were most interested in the profound, possibly allergic, response of the isolated heart to all of the nonhomologous rodent brain extracts. The responses included a high blood flow and probably caused an increased $\dot{V}_{O_2}$ in an apparently well-oxygenated organ without any evidence of increased work.

Intracerebral injections of antabolone were used to test the effect of administration of antabolene on $\dot{V}_{O_2}$ and T_b (Figs. 3 and 4) of the white rat. In an environment of 4°C. 0.1 brain equivalents* dissolved in 5 µl of artificial

---

*The mean weight of the brain of the ground squirrel is 1.4 g. A 0.1 brain equivalent is the extract from either 0.14 g of squirrel brain or 0.14 of the brain of a similar rodent, e.g., rat, guinea pig, etc.

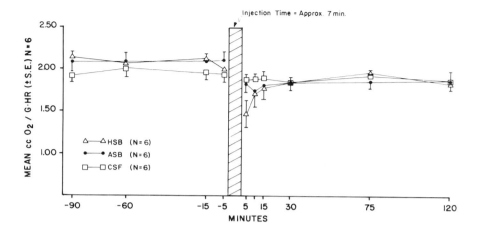

Fig. 3. Graphs of the effect on oxygen consumption in the rat following intra-cerebral injection of 0.1 brain equivalet of hibernating (HSB) and nonhi-bernating (ASB) brain extracts in 5 µl of artificial cerebrospinal fluid (CSF). CSF is the control medium.

LATERAL VENTRICLE INJECTIONS OF 0.1 BRAIN IN 5 µl CSF

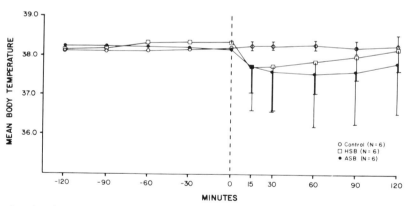

Fig. 4. Graphs of the effect on body temperature in the rat following intra-cerebral injection of 0.1 brain equivalent of hibernating (HSB) and nonhiber-nating (ASB) brain extracts in 5 µl in artificial cerebrospinal fluid (CSF). CSF is the control medium.

cerebrolspinal fluid (CSF) were injected into the lateral ventricles. The initial response was impressive. Within five minutes postinjection, the $\dot{V}_{O_2}$ fell from 2.1 to 1.25 cc O_2 g/hr, a response which is quite different than the effect of CSF alone (Fig. 3).

HSB INJECTIONS INTO THE POAH OF RATS

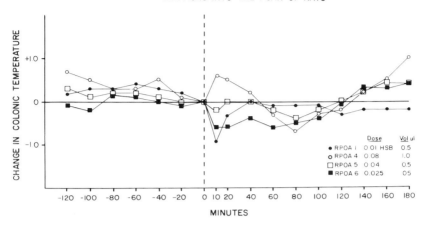

Fig. 5. The effect of varying doses of hibernating squirrel brain extract on body temperature in the rat following POAH injection.

Expectations that this technique might provide a useful bioassay in the whole animal were unfulfilled. Fig. 3 shows that the effect on $\dot{V}_{O_2}$ was not significantly different than that seen following injection of brain extracts from active squirrels. The decrease in $\dot{V}_{O_2}$ was followed by a fall in T_b (Fig. 4) of about 0.7°C using brain extracts from both hibernating and nonhibernating squirrels. This procedure at this dose level, although it showed a definite antimetabolic effect, would not distinguish hibernating from nonhibernating extracts in a bioassay.

This emphasizes the importance of dose response curves in elaborating bioassays. We have maintained that the proper control for hibernating extract is nonhibernating extract. We do not make the assumption that antabolone is absent from the active squirrel, but if it is the antimetabolic agent which induces or maintains hibernation, it should be present in greater amounts, or be in an unbound or activated state to achieve its purpose. If an assay uses a concentration at which both extracts could saturate the receptors, it would fail to be discriminatory. Fig. 5 illustrates the responses of the body temperature of the rat following intra-POAH injection of antabolone at different dose levels. This research was in collaboration with D. Caprio and M. Hawkins (unpublished).

134

In conclusion, there is strong evidence for the existance of an antimetabolic agent recoverable from the brains of hibernating animals.

SUMMARY

Brown adipose tissue does not emerge as an endocrine organ involved in the induction of hibernation. In contrast, injection of extracts from the brain of aestivating *P. aethiopicus* and of hibernating *C. tridecemlineatus* suppresses the metabolic rate and lowers the body temperature of rats. Analysis of the resting metabolism of hibernators reveals a component of NST which could be safely suppressed to induce hibernation. The need for further study in this area is stressed.

The role of serotonin and a blood-borne hibernation trigger as agents to set the endocrine stage for hibernation was reviewed. Intracerebral administration of neurotensin and bombesin in rats induces hypothermia. Bombesin is shown to reduce $\dot{V}_{O_2}$ before T_b falls.

The isolated rat heart is not a target organ of the antimetabolic hormone antabolone, but appears capable of an unusual allergic response to heterologous rodent brain extracts. Microintracerebral injections of *Citellus* brain extracts in the rat cause variable depression in $\dot{V}_{O_2}$ and T_b, but the assay does not distinguish between extracts from brains of hibernating and nonhibernating squirrels.

ACKNOWLEDGMENT

This study was aided by grant #5R01-HL21969-03 from the NHLI, Bethesda, Maryland.

REFERENCES

1. Abbotts, B., Wang, L. C. H. and Glass, J. D. (1979) Absence of evidence for a hibernation "trigger" in blood dialyzate of Richardson's ground squirrel. Cryobiology, 16, 179-183.

2. Anastasi, A., Erspamer, V. and Bucci, M. (1972) Isolation and amino acid sequence of alytsin and bombesin, two analogous active tetradecapeptides from the skin of European discoglossid frogs. Arch. Biochem. Biophys., 148, 443-448.

3. Ball, E. G. (1965) Some energy relationships in adipose tissue. Ann. N.Y. Acad. Sci., 131, 225-234.

4. Beckman, A. L. and Satinoff, E. (1972) Arousal from hibernation by intrahypothalamic injections of biogenic amines in ground squirrels. Am. J. Physiol., 222, 875-879.

5. Bigelow, W., Trimble, A. S., Schonbaum, E. and Kovats, L. (1964) A report on studies with *Marmota monax*. 1. Biochemical and pharmacological investigations of blood fat. Ann. Acad. Sci. fenn. A. 4, 71(3), 37-50.

6. Bissette, G., Nemeroff, C. B., Loosen, P. T., Prange, A. J., Jr. and Lipton, M. A. (1976) Hypothermia and intolerance to cold induced by intracisternal administration of the hypothalamic peptide neurotensin. Nature, 262, 607-609.

7. Brown, M., Rivier, J. and Vale, W. (1977) Actions of bombesin, thyrotropin releasing factor, prostaglandin E, and naloxone on thermo-regulation in the rat. Life Sci., 20, 1681-1688.

8. Brown, M., Rivier, J. and Vale, W. (1977) Bombesin: Potent effects on thermoregulation in the rat. Science, 196, 998-1000.

9. Cannon, B., Nedergaard, J., Romert, L., Sundin, U. and Svartengren, J. (1978) The biochemical mechanism of thermogenesis in brown adipose tissue, in Strategies in Cold: Natural Torpidity and Thermogenesis, Wang, L. C. H. and Hudson, J. W., eds. Academic Press, New York, San Francisco, London, pp. 567-594.

10. Carraway, R. E. and Leeman, S. E. (1973) The isolation of a new hypotensive peptide, neurotensin, from bovine hypothalami. J. Biol. Chem., 248, 6854-6861.

11. Dawe, A. R. (1978) Hibernation trigger research updated, in Strategies in Cold: Natural Torpidity and Thermogenesis, Wang, L. C. H. and Hudson, J. W., eds. Academic Press, New York, San Francisco, London pp. 541-563.

12. Dawe, A. R. and Spurrier, W. A. (1969) Hibernation induced in ground squirrels by blood transfusion. Science, 163, 298-299.

13. Dawkins, M. J. R. and Scopes, J. W. (1965) Non-shivering thermogenesis and brown adipose tissue in the human new-born infant. Nature, 206, 201-202.

14. Dawkins, M. J. R. and Hull, D. (1965) The production of heat by fat. Sci. Am. 213(2), 62-67.

15. Desautels, M. and Himms-Hagen, J. (1979) Roles of noradrenalin and protein synthesis in the cold-induced increase in purine nucleotide binding by rat brown adipose tissue mitochondria. J. Biochem. (Canada), 57, 968-976.

16. Feldberg, W. and Myers, R. D. (1976) Effects on temperature of amines injected into the cerebral ventricles. A new concept of temperature regulation. J. Physiol. (London), 173, 226-236.

17. Francesconi, R. and Mager, M. (1976) Thermoregulatory effects of monoamine potentiators and inhibitors in the rat. Am. J. Physiol., 231, 148-152.

18. Galster, W. A. (1978) Failure to initiate hibernation with blood from the hibernating Arctic ground squirrel, *Citellus undulatus*, and Eastern woodchuck, *Marmota monax*. J. Therm. Biol., 3, 93.

19. Glass, J. D. and Wang, L. C. H. (1979) Thermoregulatory effects of intracerebroventricular injection of serotonin and a monoamine oxidase inhibitor in a hibernator *Spermophilus richardsonii*. J. Thermal Biol., 4, 149-156.

20. Goldstein, A. and Lowery, P. J. (1975) The effect of the opiate antagonist naloxone on body temperature in rats. Life Sci., 17, 927-931.

21. Grubb, B. and Folk, G. E., Jr. (1977) The role of adrenoceptors in norepinephrine-stimulated V_{O_2} in muscle. Eur. J. Pharmacol., 43, 217-223.

22. Hammel, H. T., Heller, H. C. and Sharp, F. R. (1973) Probing the rostral brain stem of anesthetized, unanesthetized, and exercising dogs and of hibernating and euthermic ground squirrels. Fed. Proc., 32, 1588-1597.

23. Henshaw, R. E. (1968) Thermoregulation during hibernation: Application of Newton's law of cooling. J. Theoret. Biol., 20, 79-90.

24. Himms-Hagen, J. (1978) Biochemical aspects of non-shivering thermogenesis, in Strategies in Cold: Natural Torpidity and Thermogenesis, Wang, L. C. H. and Hudson, J. W., eds. Academic Press, New York, San Francisco, London, pp. 595-617.

25. Himms-Hagen, J., Cerf, J., Desautels, M. and Zaror-Behrens, G. (1978) in Effectors of Thermogenesis, Girardier, L. and Seydoux, J. eds. Birkhauser Verlag, Basel, pp. 119-134.

26. Hoffman, R. A. (1964) Terrestrial animals in cold: Hibernators, in Handbook of Physiology, Sec. 4, Dill, D. B., Adloph, E. F. and Wilber, C. G., eds. Waverly, Baltimore, pp. 379-403.

27. Hook, W. E. (1940) Effect of crude peanut oil extracts of brown fat on metabolism of white rat. Proc. Soc. Exp. Biol. Med., 45, 37-40.

28. Horwitz, B. A. (1978) Neurohumoral regulation of nonshivering thermogenesis in mammals, in Strategies in Cold: Natural Torpidity and Thermogenesis, Wang, L. C. H. and Hudson, J. W., eds. Academic Press, New York, San Francisco, London, pp. 619-653.

29. Hsieh, A. C. L., Pun, C. W., Li, K. M. and Ti, K. W. (1966) Circulatory and metabolic effects of control of chemical regulation of heat production. Fed. Proc., 25, 1205-1209.

30. Hudson, J. W. (1973) Torpidity in mammals, in Comparative Physiology of Thermoregulation, Vol. III, Wittow, G. C., ed. Academic Press, New York, London, pp. 97-165.

31. Hudson, J. W., and Bartholomew, G. A. (1964) Terrestrial animals in dry heat: Aestivators, in Handbook of Physiology. Adaptation to the Environment. American Physiological Society, Washington, D.C., Sec. 4, Chap. 34, pp. 541-550.

32. Hudson, J. W. and Wang, L. C. H. (1979) Hibernation: Endocrinological aspects. Ann. Rev. Physiol., 41, 287-303.

33. Jansky, L., ed. (1971) Nonshivering Thermogenesis. Proceedings of a Symposium in Prague, April 1-2, 1970. Academia, Prague, 310 pp.

34. Jansky, L. (1973) Non-shivering thermogenesis and its thermoregulatory significance. Biol. Rev., 48, 85-132.

35. Jansky, L. (1978) The sequence of physiological changes during hibernation: The significance of serotonin pathways, in Strategies in Cold: Natural Torpidity and Thermogenesis, Wang, L. C. H. and Hudson, J. W., eds. Academic Press, New York, San Francisco, London, pp. 298-326.

36. Jansky, L. and Musacchia, X. J., eds. (1976) Regulation of Depressed Metabolism and Thermogenesis. Charles C Thomas, Springfield, Ill., 276 pp.

37. Johannsen, B. W. (1960) Brown fat and its possible significance for hibernation, in Mammalian Hibernation I., Lyman, C. P. and Dawe, A. R., eds. Bull. Mus. Comp. Zool., 124, 233-248.

38. Johannsen, B. W. (1973) Effects of drugs on hibernation, in The Pharmacology of Thermoregulation, Schönbaum, E. and Lomax, P., eds. Karger, Basel, pp. 364-381.

39. Jouvet, M. (1967) The states of sleep. Sci. Am., 216(2), 62-72.

40. Kayser, Ch. (1961) The Physiology of Natural Hibernation. Pergamom, New York, Oxford, London, Paris, 325 pp.

41. Keller, A. D. (1938) Separation in the brain stem of the mechanisms of heat loss from those of heat production. J. Neurophysiol., 1, 543-557.

42. Keller, A. D. (1956) Hypothermia in the unanaesthetized poikilothermic dog, in Physiology of Induced Hypothermia, Dripps, R. D., ed. Nat. Res. Council, Washington, D.C., Publ. 451, pp. 61-79.

43. Kleitman, N. (1963) Sleep and wakefulness. Univ. Chicago Press, Chicago, 353 pp.

44. Kroll, F. W. (1952) Gibt es einen humoralen Schlafstöff im Schlafhirn? Deutsch Med. Wschr., 77, 879-880.

45. Kroll, F. W. (1953) Uber das Vorkommen von ubertragbaren schlaferzeugenden stöffen in Hirn Schlafender Tiere. Z. Ges. Neurol. Psychiat., 146, 208-218.

46. Lindberg, O. (1970) Brown Adipose Tissue. Elsevier, New York, London, Amsterdam, 337 pp.

47. Lyman, C. P. (1948) The oxygen consumption and temperature regulation of hibernating hamsters. J. Exp. Zool., 109, 55-78.

48. Lyman, C. P. (1958) Metabolism and heart rate of woodchucks entering hibernation. Am. J. Physiol., 194, 83-91.

49. Lyman, C. P. and Chatfield, P. O. (1955) Physiology of hibernation in mammals. Physiol. Rev., 35, 403-425.

50. Margulis, D. L., Goldman, B. and Finck, A. (1979) Hibernation: An opioid-dependent state? Brain Res. Bull., 4, 721-724.

51. Meeker, R. B., Myers, R. D., McCaleb, M. L., Ruwe, W. D. and Oeltgen, P. R. (1979) Suppression of feeding in the monkey by intravenous or cerebroventricular infusion of woodchuck hibernation trigger. Physiologist, 22, 86.

52. Minor, J. D., Bishop, D. A. and Badger, C. R. (1978) The golden hamster and the blood-borne hibernation trigger. Cryobiology, 15, 557-562.

53. Morrison, P. R. and Allen, W. T. (1962) Temperature response of white mice to implants of brown fat. J. Mammal., 43, 13-17.

54. Nicholls, D. G. and Heaton, G. M. (1978) The identification of the component in the inner membrane of brown adipose tissue mitochondria reponsible for regulating energy dissipation, in Effectors of Thermogenesis, Girardier, L. and Seydoux, J., eds. Experientia, Basel.

55. Nitschke, A. and Maier, E. (1932) Uber das Vergiftungabild nach Injektion von Extrakten aus lymphatischem gewebe (P-Substanz) und seine Beziehung zam Winterschlaf. A. Ges. Exp. Med., 82, 215-226.

56. Oeltgen, P. R., Bergmann, L. C. and Spurrier, W. A. (1979) Hemoglobin alterations of the 13-lined ground squirrel while in various activity states. Comp. Biochem. Physiol., 64B, 207-211.

57. Oeltgen, P. R., Spurrier, W. A. and Bergmann, L. C. (1978) Chemical characterization of a hibernation inducing trigger in plasma of hibernating woodchucks and ground squirrels. J. Thermal Biol., 3, 94-102.

58. Rivier, J. and Brown, M. R. (1978) Bombesin, bombesin analogues, and related peptides: Effects on thermoregulation. Biochemistry, 17, 1766-1771.

138

59. Rosser, S. P. and Bruce, D. S. (1978) Induction of summer hibernation in the 13-lined ground squirrel, *Citellus tridecemlineatus*. Cryobiology, 15, 113-116.

60. Smalley, R. L. and Dryer, R. L. (1967) Brown fat in hibernation, in Mammalian Hibernation III, Fisher, K. et al., eds. American Elsevier, New York, pp. 325-345.

61. Smith, R. E. and Horwitz, B. A. (1969) Brown fat and thermogenesis. Physiol. Rev., 49, 330-425.

62. South, F. E., Hartner, W. C. and Luecke, R. H. (1975) Response to preoptic temperature manipulation in the awake and hibernating marmot. Am. J. Physiol., 229, 150-160.

63. Swan, H. (1963) Kamongo and anabolone. Arch. Surg., 87, 715-716.

64. Swan, H. (1972) Comparative metabolism: Surface versus mass solved by the hibernators, in Proceedings of International Symposium on Environmental Physiology (Bioenergetics). FASEB, Bethesda, Md., pp. 25-31.

65. Swan, H. (1974) Thermoregulation and Bioenergetics. Chaps. IV, VII, XII, XIII. Elsevier, New York, Amsterdam, 430 pp.

66. Swan, H., Becker, P. and Schatte, C. (1980) Physiologic effects of brain extracts from hibernating and non-hibernating rodents on isolated perfusion rat heart. Comp. Biochem. Physiol. (In press.)

67. Swan, H., Jenkins, D. and Knox, K. (1969) Metabolic torpor in *Protopterus aethiopicus*: An antimetabolic extract from the brain. Am. Natural., 103, 247-258.

68. Swan, H. and Schatte, C. (1977) Anti-metabolic extract from the brain of the hibernating ground squirrel, *Citellus tridecemlineatus*. Science, 195, 84-85.

69. Vale, W. W., Rivier, C. and Brown, M. (1977) Regulatory peptides of the hypothalamus. Ann. Rev. Physiol., 39, 473-527.

70. Wang, L. C. H. (1978) Energetic and field aspects of mammalian torpor: The Richardson's ground squirrel, in Strategies in Cold: Natural Torpidity and Thermogenesis, Wang, L. C. H. and Hudson, J. W., eds. Academic Press, New York, San Francisco, London, pp. 109-146.

71. Wendt, C. F. (1937) Uber Wirkungen eines Extraktes aus dem braunen Fettgewebe des Winterschlafenden Igels. Z. Physiol. Chem., 249, 182.

72. Wunder, B. A., Hawkins, M. F., Avery, D. D. and Swan, H. (1980) The effects of bombesin injected into the anterior and posterior hypothalamus on body temperature and O_2 consumption. Neuropharmacology. (In press.)

73. Zirm, K. L. (1956) Ein beitrag zur Kenntnis des naturlichen winterschlafes und seines regulierenden wirkstoffes. I. Z. Naturforsch, 166, 530-538.

Published 1981 by Elsevier North Holland, Inc.
Musacchia and Jansky, eds.
Survival in the Cold
Hibernation and Other Adaptations

CHARACTERIZATION OF A HIBERNATION INDUCTION TRIGGER

PETER R. OELTGEN[1] AND WILMA A. SPURRIER[2]

[1]Lexington VA Medical Center and Department of Pathology, University of Kentucky College of Medicine, Lexington, Kentucky 40511, U.S.A.; [2]Division of Neurological Surgery, Loyola University of Chicago Stritch School of Medicine, 2160 South First Avenue, Maywood, Illinois 60153. U.S.A.

INTRODUCTION

Hibernation is defined by Hoffman [14] as a "regulated, periodic phenomenon in which body temperature becomes readjusted to new lower levels approximating ambient, and heart rate, metabolic rate and other physiological functions show corresponding reductions from which spontaneous or induced arousal to normal levels is possible at all times without the addition of environmental heat." It is one of the most striking circannual rhythmicities to be seen in mammals and is well demonstrated in the 13-lined ground squirrel (*Citellus tridecemlineatus*) and the woodchuck (*Marmota monax*). The following physiological seasonal changes are noted in these two species.

In spring and in early summer, after terminal arousing from winter hibernation, the hibernator is truly homeothermic and active. During this time reproductive activities dominate and endocrine functions are optimal. Weight increases in summer. By late fall profound changes occur involving a decline in food intake, basal metabolism, activity, and endocrine secretions. It is known that induction into hibernation can take place in both the presence or absence of food and water when the animals are appropriately exposed to cold. Ultimately in early winter, they enter a depressed state of metabolism when energy expenditure and food consumption are minimal and they resemble a poikilothermic animal. Declines in body temperature follow declines in respiration, heart rate and metabolic rate. Consequently, induction into hibernation is not due to deranged thermoregulatory mechanisms. Unfortunately, very little is known about the specific biochemical mechanisms involved in the initiation of hibernation.

The suggestion of a trigger substance which actively induces entry into the hibernating state has gained sizeable acceptance and has implicated a large number of endogenous substances from a variety of tissues such as brain [2, 28], fat [15], hormones [24], and electrolytes [23, 30]. Despite the multiplicity of hibernation induction studies, the literature provides scant experimentation designed to induce hibernation in the summertime, and hence break the circannual rhythm. Most hibernation in which the above named substances had been introduced to induce hibernation were performed during the winter hibernation season. Thus propensities for hibernation were set against the circannual backdrop which favor hibernation.

Of the large number of chemical and physiological changes known to occur during the states of activity, hibernation, and arousal, blood related studies have provided much of the data. In year-round studies Dawe and Spurrier [5] and Spurrier and Dawe [25] reported on hematologic changes in the blood of the 13-lined ground squirrel. Differences were observed in hematocrit, mean corpuscular volume, osmotic fragility of erythrocytes, folding and flattening of erythrocytes, lack of cold agglutinins or aggretation of cells, white cells, platelets and the presence of large crystalline and lipid particles in circulation. These investigators indicate that some or all of these changes may have definite advantages for free circulation of blood and exchange of oxygen to the tissues during hibernation. These changes follow a circannual rhythm in tune with the various activity states.

In another year-round study, Galster and Morrison [9] have shown that blood lipid, protein, and hematocrit levels in the 13-lined ground squirrel follow yearly cycles. Total lipid, α-lipid, β-lipid, and chylomicrons were minimal in late fall at the onset of hibernation. Moreover, proteins were minimal in late spring, except globulin which was minimal in January. A most interesting finding, which will be expanded upon, is that serum albumin concentration increased threefold (from 18 g/l) by midsummer and then to fourfold in early fall, a level which was maintained through the winter (av. 65 g/l) until spring when levels decrease sharply.

The trigger substance can induce 13-lined ground squirrels to hibernate in a cold room T_a 5°C in the summer and even when the recipient is in a warm room, i.e., T_a 23°C, with light. The trigger has not been found either in the blood of summer-active or arousing animals in winter. Thus the effective material is not described as "winter blood," but rather by the expression "hibernation blood." It is inactivated partially at 20°C, and completely at 37°C in 30 minutes in vitro.

While these previous studies have focused primarily on the physiological effects of hibernation and trigger substances, the area of chemical characterization and specific isolation of the molecule, as well as any attempts to develop a rapid and specific in vitro or in vivo assay, preferably utilizing nonhibernators, has been essentially unexplored until our recent work.

MATERIALS AND METHODS

Animal donors: Blood donors for these experiments consisted of: the wood-chucks (*Marmota monax*) and the 13-lined ground squirrels (*Citellus tridecem-lineatus*). Donor blood samples were drawn by cardiac puncture over a one-minute time span from hibernating woodchucks and ground squirrels maintained in a hibernaculum at 5°C. Plasma was separated immediately and stored at -50°C.

Primate recipients: Each of three species of macaque monkeys (*Macaca nemestrina, M. iris*, or *M. mulatta*) weighing 6-7 kg were utilized for cerebro-ventricular infusion of resolved plasma fractions derived from winter hiber-nating and summer-active woodchucks. These animals were acclimatized in special restraining chairs before experimentation and were trained to pull a lever to obtain food pellets. Surgical and other experimental methods were carried out as described previously by Myers [19] and Myers et al. [20]. Using aseptic pre-cautions, an indwelling array of four to eight guide tubes of special design was implanted chronically in the brain.

Bioassay for hibernation induction trigger (HIT): Induction of summer hib-ernation during the months of June, July and August is the assay method used to test whole plasma and fractionated plasma samples for trigger activity. A group of nontransfused animals is always maintained under similar conditions as the experimentals. These animals serve as controls and some are injected with a vehicle which is 0.9% saline. One year-old, summer-active ground squirrels, weighing 260-270 g and maintained in the warm room (23°C) are the test animals for woodchuck plasma fractions to be assayed for HIT activity. These fractions maintained at 4°C are injected into the saphenous vein of ground squirrels at a concentration of 3 mg/ml in 0.9% saline and the animals are immediately placed in the 5°C hibernaculum.

Induction of summer hibernation by the HIT-active fractions may occur as soon as two days to as long as three to five weeks after injection of the test sub-stance. All bioassays were doubleblind.

Desalting of hibernation plasma: This was accomplished using a Bio-Fiber 50 Beaker (Bio-Rad Laboratories, Richmond, California) which has a nominal molec-ular weight cut-off of 5,000.

Analytical procedures: The details of the analytical procedures of poly-acrylamide gel electrophoresis (analytical and preparative), preparative iso-tachophoresis, isoelectric focusing and affinity chromatography utilized in the characterization and isolation of the plasma fractions from hibernating and active woodchucks and ground squirrels have been described fully by Oeltgen et al. [22].

RESULTS

Isolation and partial chemical characterization of HIT from plasma of hibernating woodchucks: Prior experimental observations [7] indicated that HIT could be completely inactivated by warming the plasma from hibernating ground squirrels to 37°C for one-half hour in vitro. The thermolabile nature of this blood-borne trigger indicated to one of the investigators (Oeltgen) that it was either peptide or protein in nature. The prostaglandins, as possible candidates for trigger molecule(s), were ruled out since it is well documented that these are relatively thermostable molecules. However, three major classes of molecules or ions still remained as possible contenders for the trigger(s). These were the lipids and phospholipids which have been shown by Bragdon [3] to be the highest in hibernators of any mammalian species assayed, the catecholamines, especially those of brain origin, and small ions such as Na^+, K^+, Ca^{++} and Mg^{++}. These last two groups have been virtually ruled out by a relatively simple desalting experiment (1976) in our laboratory.

Extreme care was taken to carry out all isolation procedures of the HIT molecule at 4°C. Thirty ml of hibernating woodchuck plasma were completely desalted in a two-hour period by utilizing a hollow fiber device (Bio-Rad, Bio-Fiber 50-Beaker) which has a molecular weight cut-off of 5,000. Hence, all molecules of M.W. 5,000 and under were removed, and the lyophilized residue was then assayed for HIT activity in summer-active ground squirrels. In all cases, animals that received this desalted plasma preparation hibernated within two weeks. These results indicated the trigger was in excess of 5,000 M.W. and gave added support to the concept that it was either protein or lipid in nature.

The next step in the 1976 isolation and characterization protocol was to electrophoretically resolve the desalted plasma by the highly sensitive resolution technique of isoelectric focusing. The technique has been described by Haglund [12]. It is a procedure which is ideally suited for the analytical or preparative separation of mixed ampholytes, especially proteins, having as small a difference as 0.02 pH units at their isoelectric point (pI). Moreover, the technique is often used to determine a physical constant of a protein, its pI.

Twenty mg of the desalted plasma were electrofocused for 48 hr at 500 V in a pH gradient extending from 3.5 to 10.0 at 4°C. The results of this electrofocusing experiment are depicted in Fig. 1. Plasma was resolved into five dis-

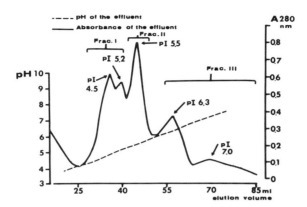

Fig. 1. Separation of hibernating woodchuck plasma by isoelectric focusing in a pH gradient from 3.5-10.0 using an LKB 8100 Electrofocusing Column. The curve with peaks shows absorption of the eluate of 280 nm. The steadily increasing curve is a plot of the pH gradient superimposed. Isoelectric point values shown at the various peaks were read from the pH gradient curve.

tinct protein fractions having pIs of 4.5, 5.2, 5.5, 6.3 and 7.0 as determined from the plot of the increasing pH curve.

To expedite the bioassay, protein components having pI values of 4.5 and 5.2 were pooled and referred to as Fraction I, while the protein component having a pI of 5.5 was termed Fraction II and the remaining two protein components having pIs of 6.3 and 7.0 were pooled and represented as Fraction III. Any lipid or phospholipid components would necessarily migrate to the anode electrode solution (comprised of phosphoric acid) because of the preponderance of negative charges associated with these molecules and thus should be completely inactivated.

These three plasma fractions were then assayed for biological activity in three groups of summer-active ground squirrels, each group being comprised of ten animals. Aliquots of Fractions I, II, and III, at a concentration of 3 mg/ml of 0.9% NaCl, were injected into the saphenous vein of ground squirrels.

All animals receiving Fractions II or III failed to hibernate, whereas eight out of ten ground squirrels receiving Fraction I hibernated within two to six days after receiving the injection. Polyacrylamide gel electrophoresis of 400 µg of Fraction I indicated that albumin was the chief protein constituent of this fraction. The major protein peak of Fraction I had a pI of 4.5 which is the physical constant previously well established for the albumin component. Moreover, this finding correlated well with the findings of Galster and Morrison [9] who noted a threefold increase in serum albumin by late summer rising to a fourfold increase in early fall, a level maintained through winter and then decreasing sharply in spring immediately following arousal. Such shifts in serum albumin correspond well with the different activity states of these hibernators. Hence, a distinct possibility exists that the HIT activity of Fraction I may be bound to or closely associated with the albumin fraction of the plasma and its physiological role may be dependent on changing albumin concentrations.

This constituted the focus of our isolation and characterization experiments as of 1976, keeping in mind the time restrictions of the bioassay for HIT activity of resolved plasma fractions. Since albumin is noted for its ability to bind a wide variety of molecules, our 1977 experimental protocol centered on attempts to isolate a completely homogeneous albumin fraction from hibernating woodchuck plasma and assay for HIT activity. To achieve this goal, three different techniques were utilized, each falling slightly short of the original objective. The ultimate goal remained resolution of a homogeneous albumin preparation.

The techniques employed in the isolation and characterization scheme for HIT active plasma fractions were: (a) Isoelectric focusing (IEF) of hibernating woodchuck plasma in a pH 3.5-10.0 gradient to obtain sufficient quantities of Fraction I which could then be further fractionated by IEF in a pH gradient extending from pH 3.5 to 6.0 [12]. (b) Affinity chromatography of both hibernating woodchuck whole plasma and of Fraction I obtained by the electrofocusing technique. The technique utilizing Affi-Gel Blue as the chromatography matrix has been thoroughly described by Travis and Panell [29] to selectively adsorb albumin from serum. (c) Preparative isotachophoresis of hibernating woodchuck whole plasma following a protocol described by Haglund [11] and Hjalmarsson [13].

The results of the 1977 fractionation and bioassay for HIT activity of resolved plasma fractions from 40 summer-active squirrels gave additional support to the concept that the HIT molecule is closely associated with or bound to albumin. An overview of these results is depicted in Fig. 2. A total of 18 of these summer-active ground squirrels hibernated. However, a much more impressive figure is that 16 out of 21 animals hibernated when injected with resolved

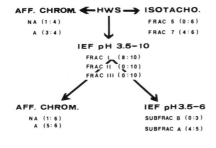

AFF. CHROM. ◄━ HWS ━► **ISOTACHO.**

NA (1:4)   FRAC 5 (0:6)
A (3:4)   FRAC 7 (4:6)

IEF pH 3.5-10
FRAC I (8:10)
FRAC II (0:10)
FRAC III (0:10)

AFF. CHROM.   **IEF pH 3.5-6**
NA (1:6)   SUBFRAC B (0:3)
A (5:6)   SUBFRAC A (4:5)

SALINE CONTROL (0:8)

Fig. 2. A summary of 1976 and 1977 bioassay results from HIT activity of resolved plasma fractions derived by affinity chromatography, isotachophoresis and isoelectric focusing. The number on the left in parentheses indicates the number of animals which hibernated, while the number on the right indicates the size of the test groups. (NA = non-albumin; A = albumin).

hibernating plasma fractions in which albumin was the predominant plasma protein. A total of eight control animals received physiological saline and none of these animals hibernated. Fig. 3 shows all resolved plasma fractions having HIT activity derived from these three techniques as compared to hibernating woodchuck whole plasma. The affinity chromatography technique has provided the most nearly homogeneous albumin fractions to exhibit HIT activity. These preparations were comprised almost entirely of albumin and a small amount of proalbumin. In a few instances, traces of transferrin were found midway in polyacrylamide gels of these fractions. The resolution of the albumin preparation has been rapid and efficient. It was used to obtain starting material for most of the experiments and will be used for future isolation studies.

Isolation studies and characterization studies--1978

Preparative gel electrophoresis: Preparative gel electrophoresis of hibernating woodchuck whole plasma or the HIT-active albumin fraction derived from the affinity chromatography column was performed utilizing the LKB 7900 Uniphor Column (LKB Instruments, Stockholm, Sweden). Three distinct protein fractions were eluted and are referred to as Fractions 1, 2, and 3. Each fraction was dialyzed and lyophilized, and 400 µg aliquots of each fraction were inspected for homogeneity by analytical PAGE as shown in Fig. 4. Bioassay results indicated that five out of five summer-active ground squirrels hibernated when injected with 3 mg/ml of Fraction 1, while three out of five hibernated when injected with Fraction 2. None of five hibernated when injected with Fraction 3

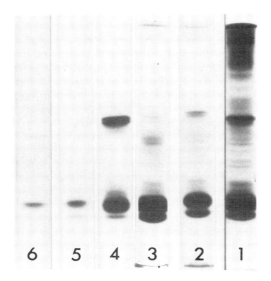

Fig. 3. Polyacrylamide gels of 400 µg aliquots obtained from all hibernating woodchuck plasma fractions exhibiting HIT activity during 1977 isolation studies. Those fractions in which albumin predominates were derived by isoelectric focusing, isotachophoresis and affinity chromatography. Gel 1 represents hibernating woodchuck whole plasma for comparison purposes. Gels 2 and 3 were derived by isoelectric focusing and represent Fraction 1 and subfraction A, respectively; Gel 4 represents Fraction 7 derived from isotachophoresis column while Gels 5 and 6 represent affinity chromatography separations of albumin fractions from hibernating woodchuck whole plasma and Fraction 1, respectively.

which contained only transferrin and thus effectively ruled out this globulin as the plasma binding protein for the HIT molecule [21].

A number of resolution techniques are being used to determine the molecular identity of Fraction I. These include preparative and analytical sodium dodecyl sulphate (SDS)-PAGE, preparative PAGE utilizing 6 M urea in the gels and chromatographic columns utilizing SDS and hydroxylapatite as the chromatography matrix.

Dose-Response Study: Summer-active ground squirrels were injected with varying concentrations of albumin fraction derived from an affinity chromatography column utilizing Affi-Gel Blue. Three dose levels (3 mg/ml, 1.5 mg/ml and 0.75 mg/ml of physiological saline) were injected into the saphenous vein of five animals in each test group. Bioassay results indicated that five out of five animals receiving the 3 mg/ml dose level hibernated, while four out of five animals receiving the 1.5 mg/ml dose hibernated, and only two out of five recipients of the 0.75 mg/ml dose hibernated. Consequently, the 3 mg/ml dose will be utilized to perform bioassay in summer-active ground squirrels in the future.

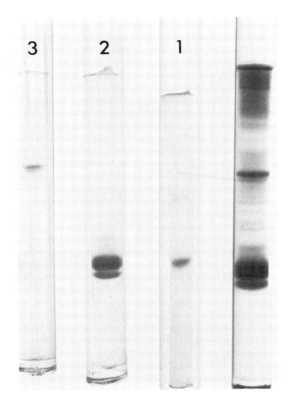

Fig. 4. Analytical PAGE of 400 μg aliquots of Fractions 1, 2, and 3 which were derived by preparative PAGE on LKB Uniphor as compared to hibernating woodchuck whole plasma.

Heat inactivation: Twenty mg of HIT-active albumin fraction derived from an affinity chromatography column were incubated in 5.5 ml of 0.9% saline at 37°C for 30 minutes. Aliquots (3 mg/ml) of this heat-treated fraction were injected immediately thereafter into summer-active ground squirrels. Of five animals receiving this heat-treated fraction, which had induced hibernation in five out of five ground squirrels in the dose response study, only one hibernated. Therefore, we concluded that inactivation could be attributed to anything other than heat. This observation is quite significant and reaffirms the contention that the HIT molecule is quite thermolabile outside of the test animal.

Protease and nuclease inactivation: If the HIT molecule is truly protein in nature as the heat-inactivation study indicates, it should be effectively destroyed by preincubation with proteases and unaffected by preincubation with

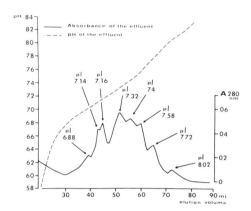

Fig. 5. Separation of 20 mg of fetal ground squirrel hemoglobin in a pH gradient extending from 1.0 to 9.0. The curve with peaks shows the absorption in the eluate at 280 nm. The steadily increasing curve is a plot of the pH gradient superimposed. Isoelectric point values shown at the various peaks were read from the pH gradient curve.

nucleases. In order to test this hypothesis, 20 mg of the albumin fraction utilized in the dose response study, were incubated at 4°C for five hours with a protease subtilisin BPN (Sigma Biochemicals, St. Louis, Mo.) at a concentration of 1 mg/ml of buffer and a sample of enzyme ratio of 100:1. None of the five recipients of this protease-treated HIT fraction hibernated. Similarly, 20 mg of the same albumin fraction utilized in the dose study was incubated overnight at 4°C with a micrococcal nuclease at a concentration of 0.002 mg/ml. Four out of five animals receiving the nuclease pretreated HIT fraction hibernated. This effectively ruled out considering nucleic acid molecules.

Based on the results of the aforementioned heat and enzyme inactivation studies, we are convinced of the thermolability and protein nature of the HIT molecule.

Development of assays for hibernation induction trigger(s): At the present time, the only fully proven assay for detection of HIT activity of resolved plasma fractions is the induction of hibernation in summer-active ground squir-

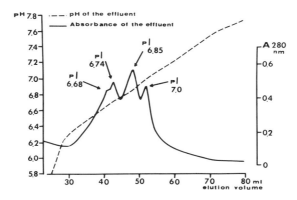

Fig. 6. Summer-Active Hemoglobin. Separation of 20 mg of summer-active ground squirrel hemoglobin by the LKB 8100 Electrofusing Column in a pH gradient extending from 7.0 to 9.0.

rels or woodchucks. This places our research involving the chemical characterization of the HIT molecule in a very restrictive time frame. Consequently, a number of attempts are now underway to develop a specific in vivo or in vitro assay for HIT which can be conducted more rapidly, during any time of the year and which preferably does not rely on induction of hibernation in summer-active animals.

Hemoglobin study alterations: Hemoglobins from fetal (Fig. 5), summer-active (Fig. 6), summer-hibernating (Fig. 7), winter-active (Fig. 8) and winter-hibernating (Fig. 9) ground squirrels were subjected to isoelectric focusing. This was done to determine whether or not significant alterations in this molecule occurred with the changing activity states and whether or not there was a reversion to a fetal type of hemoglobin in either winter or summer-hibernators. Our findings are summarized: (a) The hemoglobin molecule varies markedly with the various activity states of the hibernator. (b) Hemoglobin from summer-hibernating ground squirrels (Fig. 7) and winter-hibernating ground squirrels (Fig. 9) does not revert to a fetal type (Fig. 5) of hemoglobin. (c) A major change is noted in the number of hemoglobin peaks as these hibernators go from summer to winter season regardless of the activity state (hiber-

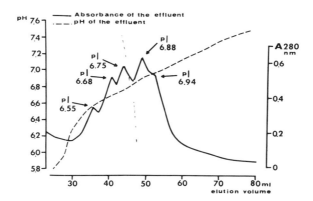

Fig. 7. Summer-Hibernating Hemoglobin. Separation of 20 mg of summer-hibernating ground squirrel hemoglobin by the LKB 8100 Electrofocusing Column in a pH gradient extending from 7.0 to 9.0.

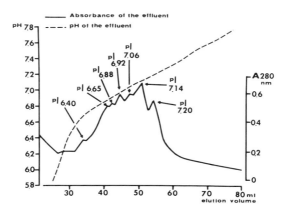

Fig. 8. Winter-Active Hemoglobin. Separation of 20 mg of winter-active ground squirrel hemoglobin by the LKB 8100 Electrofocusing Column in a pH gradient extending from 7.0 to 9.0.

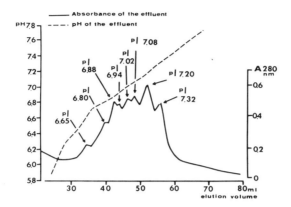

Fig. 9. Winter-Hibernating Hemoglobin. Separation of 20 mg of winter-hibernating ground squirrel hemoglobin by the LKB 8100 Electrofocusing Column in a pH gradient extending from 7.0 to 9.0.

nating or active). (d) Hemoglobin from summer-active (Fig. 6) and summer-hibernating (Fig. 7) ground squirrels have four and five hemoglobin peaks, respectively. Their elution profiles are virtually superimposable. (e) Hemoglobin from winter-active (Fig. 8) and winter-hibernating (Fig. 9) ground squirrels have seven and eight hemoglobin peaks, respectively. Their elution profiles are virtually superimposable. (f) An additional hemoglobin peak seen in summer-hibernating (Fig. 7) ground squirrels at pI 6.80 is not seen in the summer- and winter-active hemoglobins.

The possibility now exists that these additional peaks of winter-and summer-hibernating ground squirrel hemoglobins may be directly attributable to the influence of trigger(s) on the erythropoietic tissue of these hibernators and thus might constitute an assay for the HIT molecule [22].

Assay for HIT activity in monkeys: A most promising assay for detecting HIT activity of hibernating woodchuck resolved plasma fractions in primates was recently initiated by our laboratories at the VA Medical Center, Lexington, Kentucky, and the University of North Carolina, Chapel Hill, North Carolina.

This study has indicated that rodent HIT molecules can induce a number of profound physiological changes when infused into the brain ventricular fluid of monkeys. Moreover, these changes are reproducible and may thus serve as a bio-assay for HIT activity of resolved plasma fractions. Should this be the case, it would free us from the seasonally dependent bioassay currently in use.

Three test animals in this initial study were 6-8 kg female macaque and rhesus monkeys which had three guide cannulas implanted above the third brain ventricle. They also had cardiac electrodes implanted for recording heart rate and ECG and a thermistor probe for continuous monitoring of core temperature (rectal). All test substances were first dissolved in an artificial cerebrospinal fluid (CSF) developed by Myers et al. [20]. The test materials were albumin preparations derived from the affinity chromatography column of either summer-active or winter-hibernating woodchuck plasma.

In each case, preparations had been assayed for HIT activity in hibernators. A bovine serum albumin preparation also was infused into the ventricular fluid of each monkey to determine if any nonspecific protein responses could be induced.

The results of this preliminary study are summarized: (a) The heart rate decreased by 40-50% of baseline values within hours of infusion of the HIT-active albumin preparations. The response was cyclical in nature and lasted from several days to a week. (b) Core temperature declined 2°-3°C and in one animal dropped 8°C. This monkey was rewarmed with heating pads to prevent further hypothermia. (c) Visual physiological and behavioral effects were observed in all monkeys after infusion of HIT [17]. Feeding ceased in all animals when heart rates and core temperatures were declining, and resumed when they returned to baseline levels. (d) Their heads drooped and they closed their eyes when heart rate and core temperature declined, giving the appearance of an anesthetized state.

Infusion of summer-active albumin and bovine serum albumin preparations (3 mg/300 μl CSF) produced none of the aforementioned responses observed with infusion of the HIT-active albumin preparation.

DISCUSSION

Our work over the past four years, utilizing three distinct protein resolving techniques of isoelectric focusing, preparative isotachophoresis, and affinity chromatography, has given the first significant clues to the chemical identity of the HIT molecule. It is now apparent that this molecule is closely associated with albumin and that its physiological role may be dependent upon changing albumin concentrations [9]. These studies indicate that the HIT molecule is a small, thermolabile, protease-sensitive protein in excess of 5,000

M.W. However, a completely homogenous preparation which is dissociated from al-bumin and is suitable for molecular weight estimations, amino acid analyses, and sequence studies has not yet been isolated.

It is reasoned that complete chemical characterization of the HIT molecule is imperative and may have far-ranging implications. Initially, it would provide a more comprehensive understanding of the biochemical basis for the physiological phenomenon of hibernation, and ultimately, it would afford an opportunity to ex-plore the clinical potential of this molecule in nonhibernating recipients.

Some of the physiological ramifications of the HIT molecule are worth ex-amining. It is well documented that humans and other mammals that are nonhiber-nators are unable to resist the lethal effects of very low body temperature on cells and organs whereas the hibernator can survive lowering of body temperature to near zero without tissue injury or interruption of vital physiological func-tions. Evidence suggests that the HIT molecule may have a direct action on protein and/or lipoprotein synthesis thereby preparing active animals to survive extended hypothermic episodes. Studies with ground squirrel hemoglobin [22] indicate that the erythropoietic tissues of ground squirrels are responsive not only to seasonal changes but also to the direct presence of the HIT molecule and thus lend support to this hypothesis.

In our opinion the depressed metabolism noted during hibernation does not represent a reversion to an anaerobic type of metabolism. Rather, the erythro-poietic tissues respond by synthesizing hemoglobin molecules which can more readily dissociate oxygen to hypothermic tissues. This may be the role of the additional hemoglobin peaks in both natural and summer-induced hibernation in ground squirrels. It is also likely that alterations in erythrocyte membrane protein and lipoprotein are initiated by the presence of the HIT molecule. Spurrier and Dawe [25] reported a "folded-over" appearance of erythrocytes in hibernating ground squirrel blood and increased resistance of erythrocytes in osmotic fragility tests. They proposed that the folded-over erythrocytes were more pliable than normal and were thus able to circulate more readily through constricted blood vessels during hibernation. Alterations in protein synthesis, especially membrane proteins or lipoproteins, may well occur when tissues of hibernators are exposed to the HIT molecule. Thus, synthesis of proteins or lipoproteins which would maintain the flexibility of cellular membranes and prevent rapid ion shifts would be of great survival benefit for these animals during extended hypothermic periods.

In addition to these findings, our attempts at developing a more convenient bioassay for detecting HIT activity of resolved plasma fractions have given the first indications that hibernators are not unique in their ability to respond to this molecule. The clinical potential of a molecule, such as HIT, which can depress the metabolism of nonhibernators is great. For example, Goldman and Bigelow [10] have shown that rabbits injected with plasma of hibernating animals exhibited a marked lowering in the critical temperature at which ventricular fibrillation occurred. Chute [4] has demonstrated increased resistance of hibernating hosts to specific microbial, protozoan, and metazoan infections. Lyman and Fawcett [16] have shown inhibition of sarcoma development during hibernation. Musacchia and Barr [18] have demonstrated increased resistance of hibernating ground squirrels to normally lethal radiation doses, as was indicated by higher LD 50 and longer mean survival time compared to summer-active ground squirrels.

Preliminary studies utilizing three primate subjects may have provided direction to a most suitable bioassay system for testing HIT activity of resolved plasma fractions. We have noted reproducible physiological responses including hypothermia, decreased heart rate, an anesthetized appearance (Oeltgen, unpublished observations) and cessation of feeding [17]. These responses occur when the lateral brain ventricles are infused with resolved plasma fractions of hibernating woodchucks which were shown to have HIT activity in summer-active ground squirrels. The aforementioned physiological responses do not occur when similarly resolved fractions from summer-active woodchucks on bovine serum albumin are infused. Similar anorectic responses have been observed in the summer in woodchucks when they have been injected intravenously with HIT-active plasma fraction [27].

Lastly, the areas of organ preservation and transplantation in nonhibernators following infusion of the HIT molecule into the brain ventricular fluid should be thoroughly explored as well as its role in cryosurgery [1], tumor inhibition, and dietary control.

The work of Swan and Schatte [28], utilizing crude protein extracts from subcortical brain tissue of hibernating ground squirrels, raises the intriguing possibility that the HIT molecule in peripheral circulation may be similar or identical in nature to the metabolic depressant, antabolone. There is a good possibility that the HIT molecule has its origin in the brain of hibernators and in fact may be a neurohormone secreted into circulation after appropriate central nervous system stimulation. After its secretion, it may bind to circulating albumin to protect it from the destructive action of nonspecific circulating proteases. Once bound or loosely associated with albumin, the HIT molecule may influence isoenzyme, protein, and lipoprotein synthesis which alters membrane composition and permeability to ion flux to such an extent that the tissues of the hibernator or nonhibernator exposed to hypothermia can survive extended bouts of hypothermia without permanent damage.

SUMMARY

A hibernation induction trigger (HIT) is present in the plasma of hibernating woodchucks and ground squirrels which can induce hibernation when injected in summer-active ground squirrels or woodchucks. Recent biochemical characterization of the HIT molecule in our laboratory utilizing three distinct resolving techniques of isoelectric focusing, preparative isotachophoresis, and affinity chromatography clearly indicates that this molecule is bound to or closely associated with albumin in the plasma of hibernating animals. The HIT molecule is in excess of 5,000 M.W. and is inactivated by protease pretreatment or in vitro incubation at 37°C for 30 minutes. It is insensitive to nuclease pretreatment. Moreover, recent efforts to develop an assay for HIT activity of resolved plasma fractions utilizing primates as test animals has given the first indication that this molecule may initiate transient yet profound physiological and biochemical alterations in these animals. When small amounts of the HIT-active plasma fractions are infused into the brain ventricle fluid of monkeys, we observed marked physiological responses such as depressed metabolism, hypothermia, decreased heart rate and the appearance in some animals of an anesthetized state.

REFERENCES

1. Armour, D., Spurrier, W. A. and Dawe, A. R. (1974) Contractility of in situ hibernating marmot ventricle. Comp. Biochem. Physiol., 47A, 3811-3820.

2. Axelrod, L. R. (1964) Hibernation--A biological timeclock, in Progress in Biomedical Research, Vol. 10. Southwest Foundation for Research and Education, San Antonio, Texas.

3. Bragdon, J. H. (1954) Hyperlipemic and antheromatosasis in a hibernator, Citellus columbianus. Circ. Res., 2, 520-524.

4. Chute, R. (1964) Hibernation and Parasitism, in Annals Academic Scientarum Fennicae, Series A, IV, Biologica 71, Mammalian Hibernation II Symposium 1962, Suomalainen, P., ed. Annals Academiae Scientiarum Fennicae, Helsinki, p. 115.

5. Dawe, A.R. and Spurrier, W. A. (1968) Effects of hibernation and season on formed elements in the blood of 13-lined ground squirrels. Fed. Proc., 27, 2715.

6. Dawe, A. R. and Spurrier, W. A. (1968) Hibernation induced in ground squirrels by blood transfusion. Science, 163, 298-299.

7. Dawe, A. R. and Spurrier, W. A. (1971) A more specific characterization of the blood "trigger" for hibernation. (Hibernation-Hypothermia IV Symposium), Cryobiology, 8, 302.

8. Dawe, A. R. and Spurrier, W. A. (1972) The blood-borne trigger for natural mammalian hibernation in the 13-lined ground squirrel and woodchuck. Cryobiology, 9, 163-172.

9. Galster, W. A. and Morrison, P. (1966) Seasonal changes in serum lipids and proteins in the 13-lined ground squirrel. Comp. Biochem. Physiol., 18, 489-501.

10. Goldman, B. S. and Bigelow, W. G. (1964) The transfer of increased tolerance to low body temperature by heterologous transplantation of brown fat and infusions of plasma from hibernating ground hogs, in Annals Academiae Scientarum Fennicae, Series A, IV Biologica 71, Mammalian Hibernation II Symposium, 1962, Suomalainen, P., ed. Annales Academiae Scientiarum Fennicae, Helsinki, p. 175.

11. Haglund, H. (1970) Isotachophoresis--A principle for analytical and preparative separation of substances such as proteins, peptides, nucleotides, weak acids, metals. Sci. Tools, 17, 2-13.

12. Haglund, H. (1971) Isoelectric focusing in pH gradients. A technique for fractionation and characterization of ampholytes. Meth. Biochem. Anal., 19, 1-104.

13. Hjalmarsson, S. (1975) Preparative isotachophoresis--The effect of using ampholine of different pH ranges as space ions in the fractionation of serum proteins. Sci. Tools, 22, 35-38.

14. Hoffman, R. A. (1964) Terrestrial animals in cold: Hibernation, in Handbook of Physiology, Adaptation to the Environment. American Physiological Society, Washington, D.C., Sect. 4, Chapt. 24.

15. Johansson, B. (1959) Brown fat: A review. Metabolism, 8, 221-240.

16. Lyman, C. P. and Fawcett, P. W. (1954) Effect of hibernation on growth of sarcoma in hamster. Cancer Res., 14, 25-28.

17. Meeker, R. B., Meyers, R. D., McCaleb, M. L., Ruwe, W. D. and Oeltgen, P. R. (1979) Suppression of feeding in the monkey by intravenous or cerebroventricular infusion of woodchuck hibernation trigger. Physiologist, 22, 98.

18. Musacchia, X. J. and Barr, R. E. (1969) Comparative aspects of radiation resistance with depressed metabolism, in Depressed Metabolism, Musacchia, X. J. and Saunders, J., eds. American Elsevier, New York, p. 569.

19. Myers, R. D. (1977) Chronic methods--Intraventricular infusions, CSF sampling and push-pull perfusion, in Methods in Psychobiology, Myers, R. D., ed. Academic Press, New York, 281-315.

20. Myers, R. D., Yaksh, T. L., Hall, G. H. and Veale, W. L. (1971) A method for perfusion of cerebral ventricular of conscious monkey. J. Appl. Physiol., 30, 589-592.

21. Oeltgen, P. R., Bergmann, L. C. and Jones, S. B. (1978) Isolation of a hibernation inducing trigger(s) from the plasma of a hibernating woodchuck. Prep. Biochem., 8, 171-188.

22. Oeltgen, P. R., Spurrier, W. A. and Bergmann, L. C. (1979) Hemoglobin alterations of the 13-lined ground squirrel in various activity states. Comp. Biochem. Physiol., 64B, 207-211.

23. Pengelley, E. R. and Kelley, K. H. (1967) Plasma potassium and sodium concentrations in active and hibernating golden-mantled ground squirrels (*Citellus lateralis*). Comp. Biochem. Physiol., 20, 299-305.

24. Popovic, V. and Vidovic, V. (1951) Les glandes surrenales et le sommeil hibernal. Arch. Sci. Biol. (Belgrade), 3, 3-17.

25. Spurrier, W. A. and Dawe, A. R. (1973) Several blood and circulatory changes in the hibernation of the 13-lined squirrel. Comp. Biochem. Physiol., 44A, 267-282.

26. Spurrier, W. A., Folk, E. G., Jr. and Dawe, A. R. (1976) Induction of summer hibernation in the 13-lined ground squirrel by comparative serum transfusion from Artic mammals. Cryobiology, 13, 368-374.

27. Spurrier, W. A. and Oeltgen, P. R. (1980) Induction of summer hibernation and suppression of feeding in the woodchuck by infusion of hibernating woodchuck blood. Physiologist, 23, 159.

28. Swan, M. and Schatte, C. (1976) Antimetabolic extract from the brain of hibernating ground squirrels, *Citellus tridecemlineatus*. Science, 195, 84-85.

29. Travis, J. and Pannell, R. (1923) Selective removal of albumin from plasma by affinity chromatography. Clin. Chim. Acta., 49, 49.

30. Willis, J. S., Fang, L. S. T. and Foster, R. F. (1972) The significance and analysis of membrane function in hibernation, in Hibernation and Hypothermia, Perspectives and Challenges, Hannon, J. P., Willis, J. R., Pengelley, E. R. and Alpert, N. R., eds. Elsevier, Amsterdam, 123-147.

AUTORADIOGRAPHIC STUDIES OF CENTRAL NERVOUS SYSTEM FUNCTION DURING HIBERNATION

H. Craig Heller,[1] Thomas S. Kilduff,[1] and Frank R. Sharp[2]

[1]Department of Biological Sciences, Stanford University, Stanford, CA 94305, U.S.A.; [2]Department of Neurosciences, Medical Center, UCSD, San Diego, CA 92037, U.S.A.

INTRODUCTION

Research in recent years has clearly demonstrated that hibernation is a controlled, regulated phenomenon and not a reversion to poikilothermy. Studies of hypothalamic thermosensitivity show that the decline of body temperature during entrance into hibernation is regulated. The central nervous thermoregulatory system remains active during deep hibernation and can maintain a regulated differential between ambient temperature and deep body temperature. Arousal from hibernation whether partial or complete is governed by changes in hypothalamic thermosensitivity. Hence, hibernation is not a failure to regulate body temperature, but an evolutionary extension of regulated endothermy [2, 3, 5]. Many other physiological events associated with hibernation have also been shown to be regulated, for example, the timing of hibernation and reproduction, food intake and body weight [4, 10, 12, 13] and total sleep time [22].

Recent studies on a variety of species have shown shallow torpor and deep hibernation to be an extension of a more general mammalian and avian phenomenon, sleep. More specifically, hibernation appears to be an extension of slow wave sleep (SWS). Recordings of electroencephalograms and electromyograms during entrance into and exit from deep hibernation (body temperature above 25°C) and during entire bouts of shallow torpor have shown that these events occur predominantly through SWS with large decreases in wakefulness and paradoxical sleep (PS) [20, 21]. The amount of PS steadily declines as body temperature (T_b) falls so that it is essentially absent at T_bs below 26° to 28°C. Even exits from torpor, when not precipitated by external stimulation, are seen to proceed largely through SWS. During SWS in euthermic mammals and birds, there is a downward resetting of the central nervous thermoregulatory system. It is plausible to speculate that hibernating torpor, both shallow and deep, may have evolved due to the selective advantage of regulating T_b at lower and lower levels during SWS as a means of conserving energy [5, 22].

A significant difference exists, however, between resetting of the central thermoregulatory system during normal sleep cycles and during hibernating torpor. In the latter case brief episodes of wakefulness are not accompanied by a return of T_b to the euthermic wakeful level even though a decreased rate of decline or a slight rise usually occurs. This may be due to the effects of respiratory acidosis as described by Malan et al. [8] and Wünneburg and Werner [23].

The problem which challenges us in the immediate future is to understand the neural and hormonal mechanisms controlling sleep and hibernation. Our concepts and hypotheses about how brainstem systems are organized to control sleep-wakefulness are much less certain today than they were five years ago. This seems like negative progress, but much has been learned in that period [6, 9, 15]. There is much to be gained from new techniques which provide a closer look at what is happening within the central nervous system during sleep and hibernation. The present report presents one such technique and the results on its application to the study of hibernation.

The use of ^{14}C-2-deoxy-D-glucose (2DG) to measure local rates of glucose metabolism in the central nervous system and thereby to map functional pathways by their changes in rates of glucose utilization when stimulated has been described [7, 16, 17, 18, 19]. The technique is based on the fact that 2DG is taken up by cells as readily as glucose, and is believed to be phosphorylated once inside the cell but not metabolized further. Therefore, when available in trace amounts, so as not to inhibit cellular metabolism, the rate of accumulation of labelled 2DG in cells is proportional to their metabolic rate. The accumulation of 2DG can be visualized through autoradiography of sectioned tissue and measured by densitometry of the resulting autoradiographs.

EXPERIMENTAL METHODS

Animals: Golden mantled ground squirrels (*Citellus lateralis*) were prepared with chronic jugular catheters and subcutaneous thermocouple reentrant tubes. The catheters consisted of 40 mm sections of silastic tubing (.51 mm I.D. x .94 mm O.D.) which had been expanded in xylene to overlap by 4 mm a 110 mm length of polyethylene #50 tubing (.58 mm I.D. x .97 mm O.D.). The free end of the PE tubing is inserted and glued to a cavity in a Plexiglas button which is capped with a self-sealing, multiple injection, latex cap. The Plexiglas button has a basal flange which is placed subcutaneously leaving the multiple injection nipple protruding above the skin surface on the midback. The catheter is threaded subcutaneously over the shoulder to the ventral side of the neck where the silastic section is inserted into the right jugular vein so that the open end is at the level of the right atrium. The catheter is sutured to the jugular vein on the area of overlap between the silastic and the PE tubing. The jugular vein is then tied off rostral to this point. After recovering from surgery the animals were returned to a cold room (T_a = 5°C) where they resumed normal hibernation behavior of recurring bouts of torpor.

Autoradiographs: In a given experiment, an animal was transferred to a metabolism chamber (10 cm x 15 cm x 13 cm) within a larger temperature controlled unit. A thermocouple was inserted in the subcutaneous reentrant tube and a 25-gauge hypodermic needle was inserted into the latex cap of the catheter button. The needle was at one end of a length of PE 50 tubing filled with saline and leading to a 1 ml syringe outside the box. Temperatures of the animal, the air, and the metabolism chamber as well as O_2 consumption were recorded continuously. Lights were off during all experiments. The 2DG (15 microcuries/100 g body weight) was injected at a concentration of 200 microcuries/ml saline via the catheter which was then flushed with 0.3-0.4 ml saline. After the desired incubation period (45 min in euthermia, longer during hibernation) an overdose of sodium pentabarbitol was administered through the catheter.

The brain was rapidly dissected free from the skull, dropped in 2-methyl butane at -50°C, and then sectioned on an American Optical Cryocut II cryostat. The sections (20 microns thick) were mounted on glass coverslips, rapidly dried on a hot plate, and placed on x-ray film. Exposure time was seven days for euthermic controls but was adjusted to compensate for decreased metabolism and varying incubation time in the experiments done on hibernating animals. The densities of silver grains on the film overlaying 80 neural structures were quantified with a MacBeth densitometer (model TD501LB, MacBeth Inst., Newburg, N.Y.) with an aperature size of 0.43 mm. Relative densities of neural structures for each animal were determined by dividing the densitometer reading for any structure by the density of the optic tract from the same set of autoradiographs.

The optic tract is a white matter structure that has a low but significant rate of glucose utilization. Differences in dose of label, in incubation time, and in autoradiograph exposure time between different experiments should have similar percentage influences in densities of images of different structures on the autoradiographs. By expressing the densitometer readings as a ratio of structure of interest over optic tract, factors which have uniform influences over all structures but which would lead to interexperiment variability are eliminated from consideration. Data from previous 2DG experiments have been analyzed in a similar manner [16, 17].

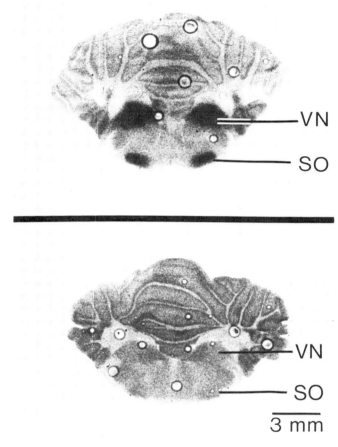

Fig. 1. Autoradiographic images of transverse sections through the cerebellum and pons of euthermic (upper) and hibernating (lower) ground squirrels injected with ^{14}C-2DG. Relative rates of glucose utilization of structures within a section are proportional to the darkness of their images. In the section of the euthermic brain, the upper pair of dark structures are the vestibular nuclei (VN) and the lower pair are the superior olivary nuclei (SO).

RESULTS AND DISCUSSION

The results are from three euthermic ground squirrels at a cold (5°C) ambient temperature, from six animals in equilibrated deep hibernation (T_a of 5°C), and from two animals arousing from deep hibernation. The euthermic animals were injected with the labelled 2DG 45 min before termination. However, since the metabolism of the hibernating animal is less than 1/30 the euthermic level, the time had to be increased. Hence, the animals in hibernation were injected 2, 4, 8, 12, 18, and 24 hrs before termination. Insufficient uptake of label occurred in the 2 and 4 hr incubations, thus results from the three longest incubation periods are presented. The animals arousing from hibernation were injected with label 15 minutes prior to stimulation of the arousal process and were then terminated at T_bs of about 35°C. The results are qualitatively illustrated by showing comparable sections from hibernating and euthermic brains, and the overall results are discussed quantitatively.

Fig. 1 shows autoradiographs of comparable transverse sections of a brain from an animal terminated while euthermic and another from an animal terminated in undisturbed deep hibernation. Both animals were at a T_a of 5°C. The sections are through the cerebellum and the pons. It is quite obvious from these two sections that the glucose utilization of the hibernating brain is much more homogeneous than in the euthermic brain. This is true throughout the brain and shall be discussed below.

Some structures, such as the vestibular nuclei and the superior olivary nuclei which are shown in these sections, undergo large reductions in relative rates of glucose utilization in hibernation in comparison to euthermia. Similarly, in Fig. 2, a comparison of transverse sections through the occipital cortex and midbrain, the inferior colliculus and the nucleus of the lateral lemniscus show large decreases in relative glucose utilization during hibernation, and layer IV of the visual cortex can not be distinguished in hibernation as it can in euthermia. In general, homogeneity of glucose utilization during hibernation is the result of large reductions in relative glucose utilization of structures which are highly active in euthermia.

There are structures, however, which show comparatively little reduction in relative glucose utilization during hibernation. It is reasonable to assume that such structures play significant functional roles during hibernation, and, therefore, are of special interest for understanding the control of hibernation. For example, Fig. 3 shows autoradiographs of transverse sections through the cerebellum and pons of euthermic and hibernating brains. Where the cerebellar and vestibular nuclei show dramatic reductions in 2DG uptake in hibernation versus euthermia, apparently there is relatively little reduction in the cochlear nucleus. Indeed, the cochlear nucleus is the most active structure in the hibernating brain. Fig. 4 presents autoradiographs of transverse sections through the cortex and the midbrain. Large reductions in 2DG uptake can be seen in the cortex and in the dorsal raphe nuclei in hibernation versus euthermia,

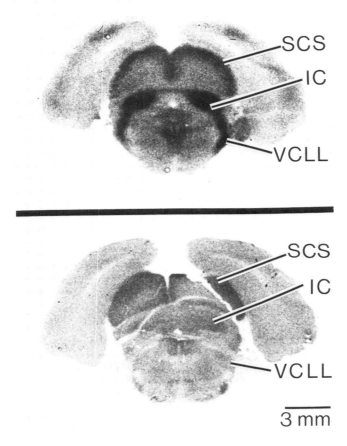

Fig. 2. Autoradiographic images of transverse sections through the occipital cortex and the midbrain of euthermic (upper) and hibernating (lower) ground squirrels injected with ^{14}C-2DG. In the euthermic brain there are relatively higher rates of glucose utilization in layer IV of the cortex, the inferior colliculus (IC), and the nucleus of the lateral lemniscus (VCLL). (The inferior colliculi are the oval-shaped dark structures in the middle of the autoradiograph from which bilaterally paired dark streaks, the nuclei of the lateral lemniscus, course ventralaterally.) In contrast, glucose utilization in comparable sections of the hibernating brain is more homogeneous.

but the superficial layer of the superior colliculus shows little reduction in activity. It is interesting to note that the structures in Figs. 3 and 4, which show little reduction in activity during hibernation, are structures which

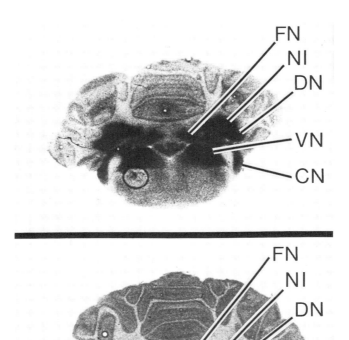

Fig. 3. Autoradiographic images of transverse sections through the cerebellum and pons of euthermic (upper) and hibernating (lower) ground squirrels injected with ^{14}C-2DG. The areas of high glucose utilization in the euthermic section are the cerebellar, vestibular (VN), cochlear (CN), fastigial (FN), dentate (DN) and interpositus (NI) nuclei. In the hibernating section the cerebellar and vestibular nuclei show a large reduction in relative glucose utilization, but the cochlear nucleus remains highly labelled.

receive sensory inputs from the periphery, the cochlear nucleus (auditory) and the superior colliculus (visual). To determine whether this observation can be extended to a generality, one must examine the data from all 80 structures examined quantitatively.

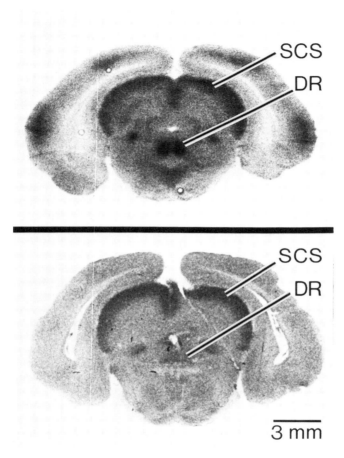

Fig. 4. Autoradiographic images of transverse sections through the cortex and midbrain of euthermic (upper) and hibernating (lower) ground squirrels injected with ^{14}C-2DG. The superior colliculus (SCS) covering the upper surface of midbrain in these sections, has a relatively high rate of glucose utilization in both euthermic and hibernating brains. Other structures, such as the dorsal raphe (DR) nuclei (paired structures near midline) and layer IV of the cortex, show large reductions in relative glucose utilization in hibernation.

The ten structures with the highest rates of glucose utilization during euthermia are ranked in the first column of Table 1. Six of these ten structures receive sensory input from the periphery, and five of these six are in the auditory system. High relative glucose utilization of auditory structures has

TABLE 1

NEURAL STRUCTURES EXHIBITING GREATEST RELATIVE GLUCOSE
UTILIZATION DURING EUTHERMIA (N=3)

Structure	$\frac{(O.D.\ structure)}{(O.D.\ optic\ tract)}$ $(\pm\ 1\ s.d.)$
Inferior colliculus	4.73 + 0.94
Vestibular nucleus	3.43 + 0.58
Superior olivary nucleus	3.34 + 0.36
Cochlear nucleus	3.28 + 0.58
Mammillary body	3.26 + 0.46
Fastigial nucleus	3.21 + 0.65
Dentate nucleus	3.18 + 0.64
Nucleus interpositus	3.07 + 0.66
Nucleus of lateral lemniscus	2.98 + 0.53
Medial geniculate body	2.95 + 0.56

also been observed in nonhibernating species, the rat and the monkey [7, 19].
On this list are the three cerebellar nuclei (fastigial, dentate, and inter-
positus) which play a major role in processing proprioceptive information.
Cerebellar nuclei also had high rates of glucose utilization in the rat [19].

The ten most active structures during undisturbed deep hibernation are listed
in Table 2 with their relative optical densities, their ordinal rankings in
hibernation and euthermia, their percent reduction in going from euthermia to
hibernation, and the significance level of the change in relative glucose utilization from euthermia to hibernation. There are two contributing factors which
place a structure on this list: (1) it can change very little with the transition to hibernation or (2) it can undergo a considerable reduction in activity
during hibernation but has been an extremely active structure in euthermia.

For example, the cochlear nucleus, the inferior colliculus, and the superior
olivary nucleus (all structures involved in processing auditory inputs) are
among the ten most active structures during euthermia, and each shows a large
and significant reduction in activity in hibernation in comparison to euthermia.
In contrast, the locus coeruleus shows only a 15.3% reduction during hibernation and therefore rises from number 52 on the list of the most active structures of the euthermic brain to number 8 on the list for the hibernating brain.
Other structures such as the superior colliculus, the dorsal tegmental nucleus,

TABLE 2

NEURAL STRUCTURES EXHIBITING GREATEST RELATIVE GLUCOSE UTILIZATION
DURING HIBERNATION AND CHANGES RELATIVE TO EUTHERMIA (N=3)

Structure	$\dfrac{\text{(O.D. structure)}}{\text{(O.D. optic tract)}}$ $\pm$ 1 s.d.	Ordinal Rank Hiber-nation	Eu-thermia	% Reduction in relative glucose util-ization from euthermic values	Signif-icance level
Cochlear nucleus	2.06 + 0.07	1	4	-37.1	.05
Lateral cuneate nucleus	1.92 + 0.20	2	45	-11.8	ns
Superior colliculus stratum super-ficiale	1.83 + 0.08	3	16	-21.9	.05
Dorsal tegmental nucleus	1.79 + 0.02	4	12	-35.7	.05
Inferior olivary nucleus	1.74 + 0.14	5	30	-27.7	.01
Inferior colliculus	1.72 + 0.11	6	1	-63.7	.05
Ventrobasal thalamus	1.72 + 0.17	7	32	-28.0	ns
Locus coeruleus	1.72 + 0.14	8	52	-15.3	ns
Nucleus of olfactory tract	1.71 + 0.04	9	23	-33.0	ns
Superior olivary nucleus	1.65 + 0.11	10	3	-49.4	.01

and the nucleus of the olfactory tract are in between these two extremes by
having fairly high relative glucose utilizations during euthermia and undergoing
small or insignificant percent reductions during hibernation.

The majority (seven out of ten) of the most active structures in the hibernating brain as in the euthermic brain are sensory, but only three auditory
structures are common to the two lists, the cochlear nucleus, the inferior colliculus, and the superior olivary nucleus. From Table 2, it is clear that the
neural basis exists for the animal in hibernation to be sensitive to environmental stimuli in multiple modalities including auditory, tactile, visual, and
olfactory. Relatively high glucose utilization in the ventrobasal thalamus may
also reflect the processing of somatosensory information. In fact, the data

TABLE 3

STRUCTURES UNDERGOING THE SMALLEST REDUCTION IN RELATIVE GLUCOSE
UTILIZATION DURING HIBERNATION IN COMPARISON TO EUTHERMIA (N=3)

	% Change*	Ordinal Rank During Euthermia	Rank During Hibernation
Anterior cerebellar hemisphere	+6.76	77	46
Paraflocculus	+4.33	76	27
Spinal white matter	+2.16	81	82
Internal capsule	+0.26	79	80
Lateral septal nucleus	-0.74	75	26
Cerebellar white matter	-2.08	80	81
Corpus callosum	-5.70	78	79
Dorsal horn of spinal cord	-10.05	73	35
Spinal trigeminal nucleus	-11.44	69	25
Lateral cuneate nucleus	-11.79	45	2
Posterior cerebellar hemisphere	-12.34	71	37
Suprachiasmatic nucleus	-12.39	61	13

*No % changes are significant

from the 80 structures reveals that no somatosensory structure undergoes a significant reduction in relative glucose utilization from euthermia to hibernation.

The three nonsensory structures which are highly active during hibernation are the dorsal tegmental nucleus, the inferior olivary nucleus, and the locus coeruleus. Little is known of the function of the inferior olivary nucleus other than it seems to serve as a major waystation for cerebellar pathways. Perhaps its high activity relates to the continued maintenance of posture during hibernation. The locus coeruleus and the dorsal tegmental nucleus may be involved in maintenance of the arousal state of the hibernator.

Structures which show little reduction in relative glucose utilization in hibernation in comparison to euthermia may be important to the hibernating state even if they are not among the most active structures. Table 3 lists 12 structures which show no significant change in relative glucose utilization with transition into hibernation. Four of these structures are less interesting because they are white matter structures with very low activities during euthermia and show no real changes in ordinal ranking with transition to hibernation.

They are spinal white matter, internal capsule, cerebellar white matter, and corpus callosum.

The remaining eight structures in Table 3 are extremely interesting because their ordinal ranks increase dramatically in hibernation in contrast to eu-thermia, and they are structures which might be expected to play important roles during hibernation. The anterior and posterior cerebellar hemispheres and the paraflocculus are involved in maintenance of background skeletal muscle tone and posture. The lateral cuneate nucleus receives direct input from proprioceptive nerve endings and relays this information to the cerebellum. A notable dif-ference between the hibernating and the hypothermic animal is the maintenance of muscle tone and the occasional adjustments of posture in the hibernating animal as opposed to the lack of muscle tone and the absence of postural control in the hypothermic animal [11, 14]. The dorsal horn of the spinal cord receives somatosensory input from the periphery including thermal afferents. The lateral septal nucleus is a limbic structure possibly involved in the interactions between the hypothalamus and brainstem systems responsible for arousal state control [2]. Last, the suprachiasmatic nucleus is well known for its role in endogenous rhythmicity which is apparently maintained intact during hibernation [1].

An interesting relationship emerges when we look at the reduction in relative glucose utilization over the 80 structures examined in going from euthermia to deep hibernation. The activity of a structure during euthermia is a strong predictor of the percent reduction in activity it will undergo during entrance into hibernation (Fig. 5). This relationship tells us two things of interest. First, there is order in the reduction of brain activity during entrance into hibernation. Such a clear correlation does not exist between percent reduction of activity during anesthesia and level of activity when unanesthetized (our analysis of data from Sokoloff et al. [19]). Perhaps this reflects the dif-ference between a natural and a forced overall reduction of brain activity.

Second, the reduction in activity during hibernation is not a direct con-sequence of temperature on the structures measured. If all gray matter structures underwent temperature dependent declines in activity with similar temperature coefficients, then the percent reduction for each structure would be the same, and the plot in Fig. 5 would have a slope of zero. There is no ready explanation for the correlation shown in this figure. One interpretation might be that level of activity in euthermia is due in part to the number of functionally active afferent inputs a structure receives. Such convergence should have a multiplicative effect on the activity of a structure. If each of

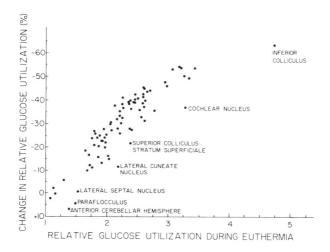

Fig. 5. A graph showing the percent change in relative glucose utilization of 80 structures of the nervous system in going from euthermia to deep hibernation. Data points are plotted as a function of the relative glucose utilization in euthermia. The structures falling farthest below the regression line are labelled.

these inputs were temperature sensitive, the percent reduction in the activity of the driven structure for a given fall in temperature would be proportional to the number of active inputs.

A third point of interest is evident from Fig. 5. Some structures have a much lower percent reduction of activity in going from euthermia to hibernation than is predicted by the regression analysis. The points corresponding to these structures fall considerably below the regression line. One might surmise that structures showing less reduction in glucose utilization may be important in the process of hibernation. The structures which fall farthest from the predicted values are those already identified as either being most active during hibernation (the inferior colliculus, the cochlear nucleus, the stratum superficiale of the superior colliculus, and the lateral cuneate nucleus) or showing the least reduction in activity during hibernation (the lateral septal nucleus, the para-

TABLE 4

NEURAL STRUCTURES EXHIBITING GREATEST INCREASES IN RELATIVE GLUCOSE
UTILIZATION DURING EXIT FROM HIBERNATION IN *C. lateralis*

Structure	Subject #59 % Change	Structure	Subject #61 % Change
Pyriform cortex	40.65	Cochlear nucleus	50.19
Basal amygdaloid nucleus	38.55	Inferior colliculus	33.72
Cingulate cortex	36.59	Interpeduncular nucleus	32.91
Medial dorsal thalamus	35.23	Oculomotor nucleus	29.83
Somatosensory cortex	33.78	Mammillary body	29.79
Vestibular nucleus	33.52	Dorsal horn of spinal cord	25.50
Frontal cortex	31.32	Basal amygdaloid nucleus	24.73
Median raphe nucleus	30.61	Medullary reticular nucleus	24.47
Medullary reticular nucleus	30.37	Solitary nucleus	24.24
Interpeduncular nucleus	29.11	Median raphe nucleus	21.95
Caudate nucleus	28.90	Dorsal raphe nucleus	19.84
Solitary nucleus	28.86	Subthalamic nucleus	19.68
Nodulus	28.40	Paratenial nucleus of thalamus	19.12
Dorsal raphe nucleus	28.00	Ventral horn of spinal cord	18.87

flocculus, and the anterior cerebellar hemisphere). It is reasonable to expect
these structures to be important to hibernation since they involve maintenance
of muscle tone and postural control (paraflocculus and anterior cerebellar
hemispheres), proprioception (lateral cuneate nucleus), possible involvement in
arousal state control (lateral septum), and processing of sensory inputs (the
inferior and superior colliculi and the cochlear nucleus).

The results concerning exit from hibernation are preliminary because they are
from only two animals which were allowed to exit fully. However, there is considerable similarity in the results from these two experiments. Table 4 lists
14 structures which showed the greatest increases in relative glucose
utilization during exit when compared to deep hibernation. Six of these
structures are common to both animals.

In a similar listing of the 12 structures showing the least increase (Table
5), seven structures were common to both animals. Among the structures showing
the greatest increase in relative glucose utilization during the exits of these
two animals, there are sensory structures (cochlear nucleus, inferior col-

TABLE 5

NEURAL STRUCTURES EXHIBITING SMALLEST INCREASES IN RELATIVE GLUCOSE
UTILIZATION DURING EXIT FROM HIBERNATION IN *C. lateralis*

Subject #59		Subject #61	
Structure	% Change	Structure	% Change
Spinal white matter	-16.86	Lateral cuneate nucleus	-24.56
Corpus callosum	-8.01	Paraflocculus	-10.20
Medullary layer of cerebellum	-4.20	Corpus callosum	-10.06
Ventral caudal nucleus of lateral lemniscus	-4.05	Medullary layer of cerebellum	-9.74
		Nucleus olfactory tract	-9.46
Superior olivary nucleus	-3.85	Internal capsule	-9.32
Internal capsule	-2.85	Superior colliculus: Stratum griseum medianem	-6.91
Pontine nuclei	-0.16		
Spinal trigeminal nucleus	4.17	Superior colliculus: Stratum griseum superficiale	-5.96
Paraflocculus	4.62	Uvula	-5.26
Nucleus olfactory tract	4.67		
Superior colliculus: Stratum griseum medianem	5.27	Lateral posterior thalamus	-4.76
		Locus coeruleus	-3.85
Suprachiasmatic nucleus	5.55	Suprachiasmatic nucleus	-3.50

liculus, vestibular nucleus, dorsal horn of the spinal cord, somatosensory cortex), but they do not dominate the lists as they did for the most active structures during euthermia and hibernation (Tables 1 and 2).

During exit one sees a large increase in some limbic system and reticular formation structures which may play important roles in the exit process, e.g., basal amygdala, dorsal and median raphe, medullary reticular nucleus, and the interpeduncular nucleus. Some motor nuclei also show great increases of activity during arousal, namely the nucleus solitarius, the oculomotor nucleus, and the ventral horn of the spinal cord.

It is interesting that there is little overlap of structures showing the most increase in relative glucose utilization during exit (Table 4) with those showing the greatest reduction in deep hibernation compared with euthermia. Because of the relationship shown in Fig. 5, the structures most active during euthermia (Table 1) are also those which show the greatest reduction in activity during hibernation. Only four structures are common to Tables 1 and 4, and three of these are sensory structures. It seems that exit from hibernation is not a direct transition from the pattern of neural function in deep hibernation to that of euthermia.   Perhaps experiments done during the actual entrance to hibernation will   reveal a transitional   pattern of   neural function and   not a direct change from euthermic to hibernating levels of   relative glucose utilization.   Comparisons of entrance and exit patterns over the same temperature ranges will be extremely interesting.

SUMMARY AND CONCLUSIONS

Conclusions must be tentative and conservative at this early stage of the study, but results are promising.   For example, they indicate the neural basis for several distinct differences between the naturally hibernating animal and the animal in forced hypothermia.   The hibernating animal remains responsive to sensory stimuli.   The results presented here show that various sensory structures retain rather high levels of glucose utilization during hibernation.   The hibernating animal maintains muscle tone and posture and our results show that the requisite neural structures retain relatively high levels of glucose utilization.   We are now engaged in comparative studies to measure the relative glucose utilizations in the brains of animals in forced hypothermia.

The ^{14}C-2DG approach is especially valuable in the study of hibernation because it provides a three-dimensional perspective on changes in central nervous system function which relate to changes in the physiology and behavior of the animal.   We believe that future results will have great significance for the understanding of central nervous system mechanisms and adaptations important in the phenomenon of hibernation.

It is reasonable to point out, however, that there are problems with the technique.   For example, there is some uncertainty about changes in glucose utilization rates translating directly into changes in neural activity.   Thus the 2DG studies will naturally lead to new electrophysiological investigations. Negative results from 2DG experiments may be difficult to interpret because they may result from neural circuits overlapping in space undergoing activity changes in opposite directions.   Or, on the motor side, very small changes in activity (hence glucose utilization) of central nervous system structures, almost undetectable by this method, may nevertheless result in very large behavioral or physiological changes due to a cascading effect.   Highly localized differences within specific structures also may be hard to interpret until we have more detailed neuroanatomical data.

With these limitations in mind, we look forward to continued interesting information coming from the application of the labelled 2DG technique to the study of hibernation.   Of immediate interest will be comparisons of undisturbed deep hibernation with forced hypothermia and with active thermoregulation during deep hibernation. Also comparisons of entrances into hibernation with exits from hibernation over similar ranges of body temperatures may reveal important mechanisms in the control of this remarkable phenomenon. The changes observed in the brains of hibernators must be viewed in the larger context of thermoregulation and sleep. Thus we are engaged in the application of this technique to euthermic animals, sleeping and waking, exposed to different peripheral and central thermal stimuli.

ACKNOWLEDGMENTS

We wish to thank the following persons for their valuable assistance in the course of this work: Charles George, Lisa Moy, Scott Sakaguchi, William Schwartz, and Nancy Thomas. The research was supported by N.I.H. grant NS10367 to H. C. Heller and U.C.S.D. Academic Senate Grant to F. R. Sharp. T. S. Kilduff was supported by N.S.F. and Danforth Foundation predoctoral fellowships.

REFERENCES

1. Cranford, J. A. (1980) Circadian rhythms of body temperature and periodic arousal in hibernating *Zapus princeps*. Cryobiology, 18, 86.

2. Heller, H. C. (1979) Hibernation: Neural aspects. Ann. Rev. Physiol., 41, 305-321.

3. Heller, H. C. and Glotzbach, S. F. (1977) Thermoregulation during sleep and hibernation. Int. Rev. Physiol., 15, 147-187.

4. Heller, H. C. and Poulson, T. L. (1970) Circannian rhythms II: Endogenous and exogenous factors controlling reproduction and hibernation in chipmunks (*Eutamias*) and ground squirrels (*Spermophilus*). Comp. Biochem. Physiol., 33, 357-383.

5. Heller, H. C., Walker, J. M., Florant, G. L., Glotzbach, S. F. and Berger, R. J. (1978) Sleep and hibernation: Electrophysiological and thermoregulatory homologies, in Strategies in Cold: Natural Torpidity and Thermogenesis, Wang, L. C. H. and Hudson, J. W., eds. Academic Press, New York, pp. 225-265.

6. Jacobs, B. L. (1978) Physiological theories of sleep: From the hypothalamus to serotonin, in Current Studies of Hypothalamic Function, Lederis, K. and Veale, W. L., eds. Karger, Basel, pp. 138-148.

7. Kennedy, C., Sakurada, O., Shinohara, M., Jehle, J. and Sokoloff, L. (1978) Local cerebral glucose utilization in the normal conscious macaque monkey. Ann. Neurol., 4, 293-302.

8. Malan, A., Rodeau, J. L. and Daull, F. (1980) Intracellular pH in hibernating hamsters. Cryobiology, 18, 100.

9. Morrison, A. R. (1979) Brainstem regulation of behavior during sleep and wakefulness. Prog. Psychobiol. Physiol. Psychol., 8, 91-113.

10. Mrosovsky, N. (1978) Circannual cycles in hibernators, in Strategies in Cold: Natural Torpidity and Thermogenesis, Wang, L. C. H. and Hudson, J. W., eds. Academic Press, New York, pp. 21-65.

11. Musacchia, X. J. (1976) Helium-cold hypothermia, an approach to depressed metabolism and thermoregulation, in Regulation of Depressed Metabolism and Thermoregulation, Jansky, L. and Musacchia, X. J., eds. Charles C Thomas, Springfield, Illinois, pp. 137-157.

12. Pengelley, E. T. and Asmundson, S. J. (1972) An analysis of the mechanisms by by which mammalian hibernators synchronize their behavioral physiology with the environment, in Hibernation-Hypothermia, Perspectives and Challenges, South, F. E., Hannon, J. P., Willis, J. R., Pengelley, E. T. and Alpert, N. R., eds. Elsevier, New York, pp. 637-662.

13. Pengelley, E. T. and Asmundson, S. J. (1974) Circannual rhythmicity in hibernating mammals, in Circannual Clocks: Annual Biological Rhythms, Pengelley, E. T., ed. Academic Press, New York, pp. 95-160.

14. Popovic, V. (1960) Physiological characteristics of rats and ground squirrels during prolonged lethargic hypothermia. Am. J. Physiol., 199, 467-471.

15. Ramm, P. (1979) The locus coeruleus, catecholamines, and REM sleep: A critical review. Behav. Neural Biol., 25, 415-448.

16. Schwartz, W. J. and Gainer, H. (1977) Suprachiasmatic nucleus: Use of ^{14}C-labelled deoxyglucose as a functional marker. Science, 197, 1089-1091.

17. Sharp, F. R. (1976) Relative cerebral glucose uptake of neuronal perikarya and neuropil determined with 2-deoxyglucose in resting and swimming rats. Brain Res., 110, 127-139.

18. Sharp, F. R., Kauer, J. S. and Shepherd, G. M. (1975) Local sites of activity-related glucose metabolism in rat olfactory bulb during olfactory stimulation. Brain Res., 98, 596-600.

19. Sokoloff, L., Reivich, M., Kennedy, C., DesRosiers, M. H., Patlak, C. S., Pettigrew, K. D., Sakurada, O. and Shinohara, M. (1977) The ^{14}C deoxyglucose method for the measurement of local cerebral glucose utilization: Theory, procedure, and normal values in the conscious and anesthetized albino rat. J. Neurochem., 28, 897-916.

20. Walker, J. M., Garber, A., Berger, R. J. and Heller, H. C. (1979) Sleep and estivation (shallow torpor): Continuous processes of energy conservation. Science, 204, 1098-1100.

21. Walker, J. M., Glotzbach, S. F., Heller, H. C. and Berger, R. J. (1977) Sleep and hibernation in ground squirrels (Citellus spp.): Electrophysiological observations. Am. J. Physiol., 233, R213-221.

22. Walker, J. M., Haskell, E. H., Berger, R. J. and Heller, H. C. (1980) Hibernation and circannual rhythms of sleep. Physiol. Zool., 53, 8-11.

23. Wünnenberg, W. and Werner, R. (1980) Effect of hypercapnia on the central regulatory system in hibernators. Cryobiology, 18, 91.

Published 1981 by Elsevier North Holland, Inc.
Musacchia and Jansky, eds.
Survival in the Cold
Hibernation and Other Adaptations

INVOLVEMENT OF NEURONAL GANGLIOSIDES AND THERMAL ADAPTATION

H. RAHMANN AND R. HILBIG
Institute of Zoology, University of Stuttgart-Hohenheim, 7 Stuttgart 70 (Hohenheim), FRG

INTRODUCTION

The activity of animals is limited to the range of body temperature within which the central nervous system remains functional. In homeothermic vertebrates this is restricted by thermoregulatory capacities. Ectothermic vertebrates also have the capacity for thermal adaptation. According to electrophysiological and biochemical data [4, 10, 11, 14], an adaptation to lowered body temperature is brought about by functional changes reflecting processes of synaptic transmission. The synapse was thought to be the primary site of adaptive changes [11]. Parallel to this, concomittant temperature dependent changes were found in poikilothermic animals in the metabolism of neuronal constituents during adaptation to cold [1, 2, 3, 6, 8, 28, 29].

Our investigations are concerned primarily with constituents of the neuronal membrane, especially gangliosides. These glycosphingolipids, containing different numbers of sialic acid (NeuAc), were found to be highly enriched in synaptic terminals [5, 13, 33], where they are located in the outer surface. Since gangliosides have the physicochemical properties to combine with Ca^{++} [16], these compounds may be involved in processes of synaptic transmission [22] and adaptive neuronal functions, including temperature adaptation [6, 17, 18, 19].

To gain more insight into the biological function of gangliosides in these phenomena, we investigated the correlations between nervous ganglioside metabolism and physiological parameters of temperature adaptation.

MATERIALS AND METHODS

The brain gangliosides of about 45 vertebrate species belonging to the classes of fish, amphibians, reptiles, birds and mammals were analyzed. Total ganglioside extracts were prepared according to Svennerholm [30], Svennerholm and Fredman [31] and Tettamanti et al. [32]. Quantitative estimations of gangliosidesialic acid were carried out according to common procedures [26]. Aliquots of purified gangliosides containing 5 µg of NeuAc were separated by thin layer chromatography on precoated silica gel plates (Fig. 1). The ganglio-

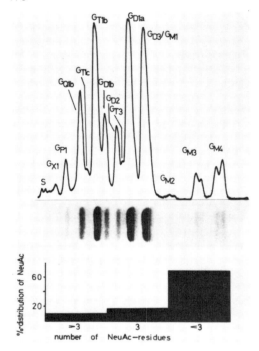

Fig. 1. Densitogram and chromatogram of gangliosides from adult chick (silica gel 60 HP-TLC plate; chloroform-methanol-water containing 0.02% MgCl-conc. ammonia = 60:36:7.5:0.4; by vol.). Lower figures show relative proportion of different gangliosides.

side spots were visualized with resorcinol reagent [30] and quantified by spectral-photometric scanning at 580mM (chromatogramspectral photometer KM 3, Zeiss). The peak areas of the densitograms (MOP II, Kontron) and the percent distribution were calculated (Hewlett Packard 9825 calculator). For comparative calculations the ganglioside fractions (mono- to pentasialogangliosides) were evaluated from their migration rates in relation to the numbers of their NeuAc-residues, respectively, and to the degree of their polarity (migration rate slower or faster than the trisialogangliosides). Following increasing migration rates, the ganglioside fractions were numbered and identified, as far as possible, in comparison with standards and named according to Svennerholm's nomenclature [31].

RESULTS

Phylogenetic differences in brain gangliosides: Comparative studies of brain gangliosides from a large number of vertebrate species reveal [7] that the ganglioside composition of species belonging to different systematic groups correlates to the level of nervous tissue organization (Fig. 2). In higher vertebrates, mammals and birds, ganglioside concentration is in the range of 500 to 1000 µg ganglioside bound NeuAc per gr. fresh wt. In contrast, concentration in the lower vertebrates, reptiles, amphibians and fish, is about 110 to 500 µg.

Besides these differences, there exists a remarkable variability in ganglioside composition. In the brain of adult mammals and birds the ganglioside pattern consists mainly of mono- and disialogangliosides (60-70%). The content of trisialogangliosides is about 20-30% and that of the more polar fractions only 10-15%. In reptilia the ganglioside composition changes in favour of tri- and tetrasialogangliosides, and the whole ganglioside pattern is relatively less polar. In amphibia and in most fish there exists a preponderance of the polar ganglioside fractions (up to 60%). It is noteworthy that the variability of the brain ganglioside pattern is distinct within the classes of lower vertebrates, especially in teleost fish (Fig. 2).

Adaptation of brain gangliosides in species living at different ambient temperatures: When comparing the data cited, it must be emphasized that no clear correlations (especially between the ganglioside pattern and the systematic position of the species investigated) could be followed. Therefore the question was raised, are the differences in the ganglioside concentration and pattern correlated to the environmental temperature? Recall that the lower vertebrates (fish, amphibians and reptiles) are poikilotherms, while birds and mammals are homeotherms with the capacity for temperature regulation. The significantly higher amount of brain gangliosides and the preponderance of less polar fractions (Fig. 2) are considered to be an expression of adaptive evolutionary processes in which thermoregulation and homeostasis of physiological functions are deeply involved.

To elucidate this hypothesis, an investigation was made of brain ganglioside patterns in poikilothermic vertebrates of species belonging to the same taxonomic category, e.g., teleost fish, but adapted to different and extreme temperature biotopes [21]. The brain gangliosides of a shallow water ice fish (*Trematomus hansoni*) from the south polar regions (water temperature of 1.5°C) were compared with those of the eurythermic codfish (*Gadus morhua*), the European whiting (*Merlangius merlangus*), and the partly heterothermic tunafish (*Thunnus thunnus*). The well-known pattern of the cortex from a rabbit served as a standard. Fig. 3 shows that the ice fish brain ganglioside pattern is the most

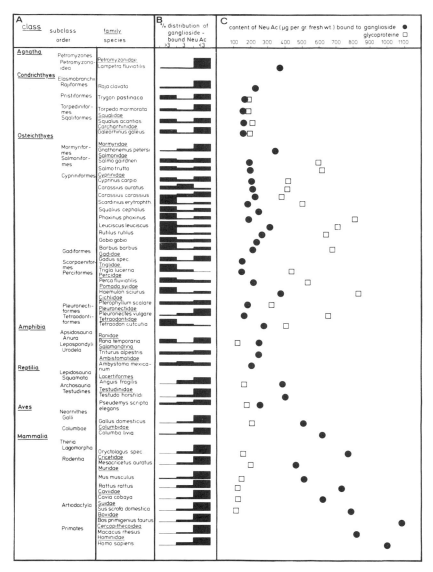

Fig. 2. Brain gangliosides in vertebrates including taxonomical classification of species (A), relative proportion of main ganglioside fraction groups migrating slower or faster than trisialogangliosides (B), content of ganglioside bound NeuAc (C), content of NueAc bound to ganglioside and glycoprotein.

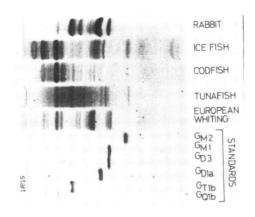

Fig. 3. Thin layer chromatogram of brain gangliosides from teleost fishes adapted to biotopes of extremely different temperatures; several ganglioside standards are included.

polar. In all other fish and higher vertebrates the highest polar ganglioside fraction is a G_{p1}. Four additional fractions could be detected in the ice fish. These are more highly sialylated than G_{p1} and account for about 45% of all ganglioside species. These data indicate that the neuronal membranes of the ice fish, according to their utmost sialylation, are extremely polar. This polysialylation can be taken as an adaptation mechanism in polar fishes, in addition to other well known mechanisms (e.g., lack of hemoglobin and antifreeze proteins), that enable these animals to keep their neuronal membranes functional in the extreme low temperature of the Antarctic Sea.

<u>Acclimatization in poikilothermic vertebrates</u>: The composition of brain gangliosides had been examined in various species in terms of adaptation to seasonally induced fluctuations of the environmental temperature. It is reasoned that as a result of long-term adaptations to lowered temperatures a reorganization in this feature of brain gangliosides evolved resulting in a more polar pattern. This polysialylation is indicated by a decrease in the mono- and disialogangliosides, while tri- and especially the polar tetra- and pentasialogangliosides increased. Modified types of such a kind of polysialylation of brain gangliosides have been observed in several species (trout, goldfish, carp, plaice) as a function of a seasonal adaptation to lowered temperatures (Fig. 4).

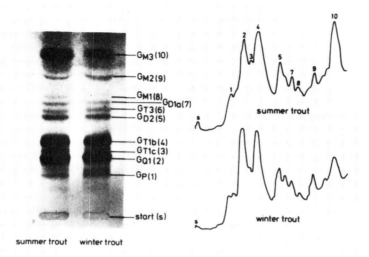

difference percent changes of single brain ganglioside fractions

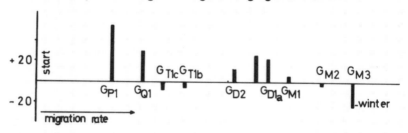

Fig. 4. Thin layer chromatogram and densitogram of brain ganglioside patterns from rainbow trout acclimatized to summer (19°-22°C), and winter (4°-6°C) conditions. Lower figure shows difference percentage of single ganglioside fractions of winter in comparison to summer animals (= base line).

<u>Acclimation in poikilothermic vertebrates</u>: From the data above it cannot be concluded that the phenomenon of polysialylation of brain gangliosides in teleost fish during seasonal acclimatization is either a single adaptive reaction of nervous tissue that compensates for environmental temperature changes or is a summation of different processes. Therefore it was necessary to investigate whether these changes in ganglioside synthesis might also occur under experimental laboratory conditions.

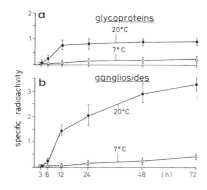

Fig. 5. Incorporation of labelled (N-[3H]-acetylmannosamine) in glycoproteins and gangliosides in the brains of goldfish adapted for three weeks to 20°C and 7°C, respectively.

Different experimental approaches were taken. One was to investigate changes in the content and composition of brain gangliosides after transition of the animals from warm to cold water following long-term acclimation under defined conditions. The other was to analyze ganglioside metabolism by means of radioactive tracer studies [26] after changed temperatures.

The application of N-[3H]-acetylmannosamine (as a precursor for NeuAc) into goldfish brains after a three-week preadaptation to 20° and 7°C, respectively, and incorporation at periods between 3 and 72 hrs reveals that the ganglioside metabolism is much more sensitive to temperature changes than glycoprotein metabolism (Fig. 5). Gangliosides are labelled four to five times higher in warm adapted fish than in cold ones. The sialoglycoprotein metabolism is only two to three times as intensive [26]. Additionally, it was shown that at lower temperatures the turnover of the polar multisialogangliosides was diminished to a greater extent than that of oligosialogangliosides. Consequently during long-term adaptation this change in metabolism causes a polysialylation of these membrane specific brain lipids. This response has been shown for various species [2, 6, 18].

Changes of brain gangliosides during heterothermic phases in vertebrates with thermoregulation

Perinatal development: In adult homeothermic vertebrates, birds and mammals, the ability to tolerate large changes in body temperature is poorly developed.

184

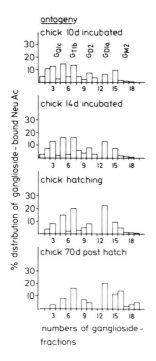

Fig. 6. Percent distribution of neuronal ganglioside bound NeuAc during ontogeny of chick from ten days incubation time to 70 days posthatch.

However, during early neonatal development, a heterothermic phase occurs during which the newborn animal is not yet able to keep body temperature constant. While there are several developmental profiles of brain ganglioside patterns for different vertebrates [12, 25, 27], there are only a few studies which provide evidence for the assumption that, during early development of homeothermic vertebrates, the differentiation of brain gangliosides is correlated to thermobiological parameters [9, 18, 27].

It has been shown in the chicken (Fig. 6) that simultaneous to gradual acquisition of thermoregulation and resistance of cold stress after hatching, the composition of brain gangliosides changes significantly [25]. This critical developmental period is characterized by an increase of the disialoganglioside G_{D1a} and a decrease of the higher sialylated tetra- and especially pentasialogangliosides. Molecular shift of brain gangliosides is finished at about 12 days after hatching, a time when temperature regulation becomes established. Similar results were obtained in ontogenetic studies with mice and rats [20, 27]. These data obviously reflect an important and more general mechanism of modulating the thermosensitivity of neuronal membrane mediated processes.

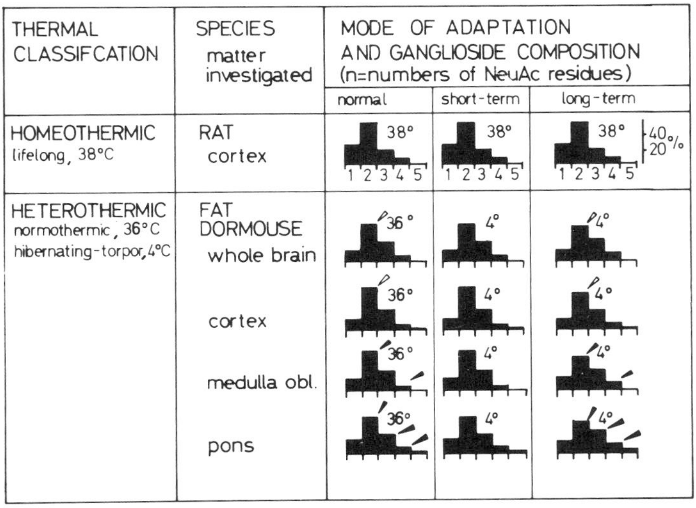

THERMAL CLASSIFCATION	SPECIES matter investigated	MODE OF ADAPTATION AND GANGLIOSIDE COMPOSITION (n=numbers of NeuAc residues)		
		normal	short-term	long-term
HOMEOTHERMIC lifelong, 38°C	RAT cortex	38° 1 2 3 4 5	38° 1 2 3 4 5	38° ⌐40% ⌐20% 1 2 3 4 5
HETEROTHERMIC normothermic, 36°C hibernating-torpor,4°C	FAT DORMOUSE whole brain	36°	4°	4°
	cortex	36°	4°	4°
	medulla obl.	36°	4°	4°
	pons	36°	4°	4°

Fig. 7. Percent distribution of brain gangliosides according to their polarity (mono- to pentasialogangliosides) in a hibernator (fat dormouse) and rat cortex (= control). During hibernation there are significant (open darts) and highly significant (closed darts) changes in ganglioside pattern, causing a polysialylation of neuronal membranes.

Hibernation: The brain ganglioside composition of homeothermic vertebrates (e.g., rat) and of hibernators (e.g., fat dormouse, *Glis glis*), either in their normothermic or heterothermic phase, differs significantly (Fig. 7). When analyzing the whole brain pattern of the fat dormouse during normothermic and torpor phases, a polysialylation can be noted. This is due to a decrease in the disialoganglioside fractions.

The gangliosides from different brain parts (cortex, medulla and pons) were investigated in detail and showed regional variations. A significant polysialylation effect was registered in hibernating dormice, especially in the medulla and pons, the two regions which are functional during torpor. These effects in the forebrain and cortex were less distinct.

Results indicating a similar tendency were obtained from golden hamsters and dsungarian hamsters, two species which show less pronounced hibernation characteristics [6].

CONCLUSIONS

The present results provide evidence for a general phenomenom in the CNS of

thermal classi-fication	species	adaptation of ganglioside composition		NeuAc >3 3 <3	neuronal activity		
		normal	short-term	long-term	normal	short-term	long-term
poikilothermic							
stenothermic,cold (-2 to +4°C)	ice fish (Trematomus hansoni)	±0°	>4°	>4°	+++	-	-
stenothermic,warm (+22 to +36°C)	tropic fish (Tilapia mariae)	25°	<18°	<18°	+++	-	-
eurythermic (4 to 28°C)	goldfish (Carassius auratus)	25°	5° 0	5°	+++	-	++
homeothermic							
lifelong (38°C)	adult rat (Rattus norvegicus)	const.	const.	const.	+++	+++	+++
heterothermic							
ontogenetic (4 to 38°C)	neonatal rat (Rattus norvegicus)	37°	diff.	37°	+	++	+++
hibernation (4 to 38°C)	golden hamster (Mesocricetus auratus)	37°	5°	5°	+++	+	+
	fat dormouse (Glis glis)	36°	4°	4°	+++	+	+

Fig. 8. Percentage of main brain ganglioside fraction groups (migrating slower or faster on TLC as the trisialogangliosides) and neuronal activity of different vertebrates according to thermal classification and ability to adapt to changes in environmental temperature.

vertebrates, which shows that a definite correlation exists in the ability of organisms to adapt to changes in environmental temperature and (a) with the polarity of neuronal membranes and (b) with ganglioside composition and concentration. The ganglioside compositions, which can show extreme variation in the brain of different vertebrates, range from less polar oligosialogangliosides to polar multisialogangliosides. There is a consistency in those animals adapted to constant temperature conditions: Antarctic ice fish to about 0°C, tropic fish to 25°-30°C or homeothermic mammals or birds to 37°-39°C (Fig. 8). These groups of animals cannot tolerate wider changes in body temperature, probably because their neuronal gangliosides are highly specialized with respect to the neuronal membranes' transmission abilities. In contrast, a poly- or desialylation of neuronal gangliosides was observed in those vertebrates which are able to tolerate larger changes in body temperature. Consequently, in order to survive, eurythermic species, (goldfish, carp, plaice), homeotherms in their heterothermic phases during ontogeny (neonatal chick, mouse, rat), and heterothermic mammals during hibernation (golden hamster, fat dormouse) are able to adapt their metabolism to variations in body temperature caused by environmental thermal fluctuations (Fig. 8).

Recent physicochemical approaches show that polysialylation of brain gangliosides at lowered temperatures induces changes in the affinity of gangliosides to complex with Ca^{++}. These ions probably cause the changes in fluidity of neuronal membranes [15, 16, 19]. These results correspond to electrophysiological and behavioral findings. Adaptive long-term changes in brain ganglioside composition of teleost fish occur simultaneously with the return of postsynaptic amplitudes of photic evoked potentials and conduction [19, 23, 24].

These phenomena reflect a mechanism for modulating the thermosensitivity of membrane-mediated processes of transmission.

SUMMARY

The concentration and composition of brain gangliosides, from lower to higher vertebrates (fishes to mammals) show remarkable variations such as increase in amount of total gangliosides and decrease in number of single fractions. Considerations of the ecofactor temperature provide correlations between systematic position of species, thermal adaptation status and brain ganglioside composition. Lowering the environmental temperature induces a long-term formation of more polar polysialogangliosides. This general tendency has been demonstrated:

(1) for fishes living at different thermal biotopes (Antarctic ice fish versus Mediterranean ones);

(2) for fishes during the phase of seasonal acclimatization (trout, carp, goldfish, plaice);

(3) for fishes during experimentally induced acclimation (trout, carp);

(4) for homeotherms during the heterothermic phase of perinatal development (chick, mouse, rat); and

(5) for hibernators (fat dormouse, golden hamster). During hibernation, polysialylation of neuronal membrane bound gangliosides takes place in midbrain, medulla and pons but not in forebrain or cerebellum.

REFERENCES

1. Breer, H. (1975) Ganglioside pattern and thermal tolerance of fish species. Life Sci., 16, 1459-1463.

2. Breer, H. and Rahmann, H. (1976) Involvement of brain gangliosides in temperature adaptation of fish. J. Therm. Biol., 1, 233-235.

3. Cossins, A. R. (1977) Adaptation of biological membranes to temperature. The effect of temperature acclimation of goldfish upon the viscosity of synaptosomal membranes. Biochim. Biophys. Acta, 470, 395-411.

4. Goldman, S. S. (1975) Cold resistance of the brain during hibernation. III. Evidence of a lipid adaptation. Am. J. Physiol., 228, 834-838.

5. Hansson, H. A., Holmgren, J. and Svennerholm, L. (1977) Ultrastructural localization of cell membrane GM1 ganglioside by cholera toxin. Proc. Nat. Acad. Sci., 74, 3782-3786.

6. Hilbig, R. and Rahmann, H. (1979) Changes in brain ganglioside composition of normothermic and hibernating golden hamsters (*Mesocricetus auratus*). Comp. Biochem. Physiol., 62B, 527-531.

7. Hilbig, R. and Rahmann, H. (1980) Variability in brain gangliosides of fishes. J. Neurochem., 34, 236-240.

8. Hilbig, R., Rahmann, H. and Rösner, H. (1979) Brain gangliosides and temperature adaptation in eury- and stenothermic teleost fishes (carp and rainbow trout). J. Therm. Biol., 4, 29-34.

9. Hilbig, R., Segler, K., Rösner, H. and Rahmann, H. (1979) Changes in brain gangliosides in homeothermic mammals during heterothermic phases. Hoppe-Seyler's Z. Physiol. Chem., 360, 282.

10. Konishi, J. and Hickman, C. P. (1964) Temperature acclimation in the central nervous system of rainbow trout (*Salmo gairdnerii*). Comp. Biochem. Physiol., 13, 433-442.

11. Lagerspetz, K. Y. H. (1974) Temperature acclimation and the nervous system. Biol. Rev., 49, 477-514.

12. Merat, A. and Dickerson, J. W. T. (1973) The effect of development on the gangliosides of rat and pig brain. J. Neurochem., 20, 873-880.

13. Morgan, J. G., Zanetta, J. P., Breckenridge, W. C., Vincendon, G. and Gombos, G. (1973) The chemical structure of synaptic membranes. Brain Res., 62, 405-411.

14. Musacchia, X. J., Ackerman, P. D. and Entenman, C. (1976) In vitro oxidation of [3-14C] acetoacetate in brain cortex from hibernating hamsters and ground squirrels. Comp. Biochem. Physiol., 55B, 359-362.

15. Probst, W. and Rahmann, H. (1980) Influence of temperature changes on the complexation ability of gangliosides with Ca2+. J. Therm. Biol. (In press.)

16. Probst, W., Rösner, H., Wiegandt, H. and Rahmann, H. (1979) Das Komplexationsvermogen von Gangliosiden für Ca2+. I. Einfluss mono- und divalenter Kationen sowie von Acetylcholin. Z. Physiol. Chem., 360, 979-986.

17. Rahmann, H. (1976) Possible functional role of gangliosides, in Ganglioside Function: Biochemical and Pharmacological Implications, Porcellati, G., Ceccarelli, B., and Tettamanti, G., eds. Plenum, New York, p. 151.

18. Rahmann, H. (1978) Gangliosides and thermal adaptation in vertebrates. Japan. J. Exp. Med., 48, 85-96.

19. Rahmann, H. (1980) Gangliosides and thermal adaptation. Adv. Exper. Med. Biol., 125, 505-514.

20. Rahmann, H. and Hilbig, R. (1979) Gangliosides in the CNS of senescent rats, in Glycoconjugates, Schauer, R., Boer, P., Buddecke, E., Kramer, M. F., Vliegenthart, J. F. G., and Wiegandt, H., eds. Thieme, Stuttgart, p. 672.

21. Rahmann, H. and Hilbig, R. (1980) Brain gangliosides are involved in the adaptation of Antarctic fish to extreme low temperatures. Naturwiss., 61, 259.

22. Rahmann, H., Rösner, H. and Breer, H. (1976) A functional model of sialo-glyco-macromolecules in synaptic transmission and memory formation. J. Theor. Biol., 57, 231-237.

23. Rahmann, H., Schmidt, W. and Schmidt, B. (1980) Influence of long-term thermal acclimation on the conditionability of fish. J. Therm. Biol., 5, 11-16.

24. Reckhaus, W. and Rahmann, H. (1979) Long-term thermal adaptation of evoked potentials in the optic tectum of goldfish. JRCS Med. Sci., 7, 290.

25. Rösner, H. (1979) Changes in the contents of gangliosides and glyco-proteins in the ganglioside pattern of the chicken brain. J. Neurochem., 24, 815-816.

26. Rösner, H., Breer, H., Hilbig, R. and Rahmann, H. (1979) Temperature effects on the incorporation of sialic acid into gangliosides and glycopro-teins of the fish brain. J. Therm. Biol., 4, 69-73.

27. Rösner, H., Segler, K. and Rahmann, H. (1979) Changes of brain ganglio-sides in chicken and mice during heterothermic development. J. Therm. Biol., 4, 121-124.

28. Roots, B. I. (1979) Phospholipids of goldfish (Carassius auratus L.) brain: The influence of environmental temperature. Comp. Biochem. Physiol., 25, 457-466.

29. Selivonchick, D. P., Johnston, P. V. and Roots, B. J. (1977) Acyl and alkenyl group composition of brain subcellular fractions of goldfish (Carassius auratus L.) acclimated to different environmental temperatures. Neuro-chem. Res., 2, 379-393.

30. Svennerholm, L. (1957) Quantitative estimation of sialic acids. II. A colorimetric resorcinol-hydrochloric acid method. Biochim. Biophys. Acta, 24, 604-611.

31. Svennerholm, L. and Fredman, P. (1980) A procedure for the quantitative isolation of brain gangliosides. Biochim. Biophys. Acta, 617, 97-109.

32. Tettamanti, G., Bonali, F., Marchesini, S. and Zambotti, V. (1973) A new procedure for the extraction, purification and fractioning of brain ganglio-sides. Biochim. Biophys. Acta., 296, 160-170.

33. Wiegandt, H. (1967) The subcellular localization of gangliosides in the brain. J. Neurochem., 14, 671-674.

LIFE AT LOW AND CHANGING TEMPERATURES: MOLECULAR ASPECTS

H. W. BEHRISCH, D. H. SMULLIN AND G. A. MORSE
Institute of Arctic Biology, University of Alaska, Fairbanks, AK 99701, U.S.A.

INTRODUCTION

There is much research that suggests that low temperature, within certain limits, is not a major environmental disturbance to an organism. Rather, it appears that thermal change affects the animal in such a way that adaptive responses become necessary. The major disturbance elicited by a change in ambient temperature occurs at the level of regulation of intermediary metabolism and the intracellular milieu. Any intracellular adaptation to temperature must also reflect changes in the intracellular environment that are brought about by a change in temperature.

Adaptations of glycolytic enzymes from various tissues of the hibernator, as well as hetero- and homeothermic tissues of boreal mammals, involve changes in enzyme activity and certain subtle, but crucial adjustments in regulation. Laboratory exposure to chronic cold (-40°C) of the varying hare (*Lepus americanus*) results in a widespread change in the isoenzyme profile of glycolysis in the ear, but a rather modest alteration in activity. As expected, the same enzymes from the deep tissues remain unaltered. The same tissues from wild hares taken in winter show this isoenzyme shift but also show a dramatic rise in enzyme activity, and presumably, in glycolytic capacity. Thus simple studies on temperature adaptation tend to obscure factors that, in concert with temperature, set in motion the animal's adaptive machinery.

In the hibernator, *Citellus undulatus*, there appears a striking dichotomy among the animal's tissues as it adapts throughout the hibernation cycle. A number of tissues and their pathways remain active throughout the year while others do not. In the former category, changes in the isoenzyme profile occur in anticipation of a change from one state to the other, while in those tissues that are relatively functionless, during one state or another, no such alterations in isoenzyme profile are observed. The isoenzyme changes are such that precision and flexibility of metabolic control are retained, allowing the animal to take advantage of the great drop in body temperature to reduce metabolism. The absence of an isoenzyme shift in tissues that are relatively inactive during either hibernation or nonhibernation, combined with a marked drop in enzyme activity during hibernation, allows the animal to use one format of the enzyme for specific tissues. This provides a considerable metabolic economy in adapting to a changing thermal regime throughout the year.

Most organisms live within relatively narrow confines of internal temperature. In alpine or near arctic regions however, where seasonal or even daily thermal fluctuations are marked, life persists and flourishes. Animals in these regions utilize a variety of mechanisms as solutions to thermal stress.

A number of cold-blooded animals become torpid and overwinter with body temperatures just above freezing. Others, notably the insects, actually tolerate freezing, when their natural cryoprotective agents no longer suffice to prevent the formation of ice [29]. In the classically homeothermic birds and mammals, the picture is equally diverse. The extremities of many mammals are supplied by a circulatory arrangement that functions to reduce heat loss through these exposed areas of the body. As a result, the tissues become heat regulators and temperatures within them vary widely, particularly during winter. It is clear that such adaptations to winter life involve the expenditure of large amounts of energy which must be constantly obtained and consumed, often at suboptimal temperatures.

Another solution to the cold problem is typified by the hibernator, which largely avoids the rigors of winter. However, the hibernator can only do this because of an extremely active biosynthetic machinery in the active season and a tight husbanding of finite metabolic reserves during hibernation.

This report will be limited to a selected group of northern animals with which we have been working for the last few years. We will focus primarily on mammals with heterothermous extremities and the hibernator, and some of the enzymatic mechanisms which they appear to utilize in adapting to cold.

One of the basic problems of the physiologist studying adaptations to changing temperature is that many of the mechanisms regulating intermediary metabolism appear to be directly affected by temperature. In bacterial and mammalian enzyme systems, large and often divergent effects of temperature on enzyme modulation and enzyme substrate associations can occur [17, 42]. From those studies it was argued that if the changes observed in homeothermic enzymes were extended to heterothermic animals, they would be incompatible with survival in a hypothermic situation.

For example, the very large temperature effects on the affinity of rabbit liver fructose bisphosphatase, a rate-limiting enzyme of gluconeogenesis, would totally shut off the pathway in hibernating animals such as the Arctic ground squirrel (*Citellus undulatus*). In this connection, it appears that the pathway is operative and indeed favored during hibernation [15, 24]. This is supported by data on the intact organism and by kinetic data on the isolated enzyme from liver [5]. From such a background, a large literature has accumulated on the kinetic adaptations to low temperature in ectotherms [16, 18, 19, 41].

The central finding of most of those studies was that as temperature decreases, the affinity of an enzyme for its substrate increases, as reflected by the apparent Michaelis constant, or $S_{0.5}$ value. It was argued that this enhanced enzyme substrate interaction would tend to hold enzyme activity essentially independent of thermal change and in some cases even enhance it at low temperature. An often observed characteristic of such results was a U-shaped Km-temperature curve, the bottom of which (maximum affinity for substrate) approached the temperature of acclimatization [3, 18, 19].

The ensuing argument on strategies of thermal adaptation assumed that concentrations of the substrate themselves are also reduced with a fall in temperature, which in fact does not appear to be the case. The major feature of an effect of a drop in temperature is a fall in the turnover rate of the intermediates which are maintained, within limits, at certain steady state concentrations. From this it follows that a large increase in enzyme substrate affinity would be nonadaptive since an $S_{0.5}$ value which is far below the physiological concentration of an intermediate is as valueless to the cell as an $S_{0.5}$ value which far exceeds the concentration of that intermediate.

If the concept of enzyme substrate affinity is to have any relevance in current theory of enzyme regulation, then $S_{0.5}$ values must approach the physiological concentrations of the various metabolites. Detailed considerations of metabolic regulation are presented by Atkinson [2].

In our view, a major shortcoming of these enzyme studies is that they were invariably conducted at constant pH values over a wide temperature range. From the now classic studies on the effects of temperature on pH of physiological fluids [26, 30, 37, 38], it is evident that under such conditions the environment of the enzyme becomes more acidic as the temperature is lowered, a situation generally not observed in the physiological state. Homeostasis of the cell's acid-base relationship is of critical selective importance and a number of intracellular buffer systems exist to maintain the pH of the intracellular milieu at a point of relative neutrality, whatever the temperature of the system [28, 35, 36, 43, 44].

The same experiments described above can be performed with one change: correction of the pH-temperature effect. In such a situation, the affinity of lactate dehydrogenase from heterothermous skeletal muscle of the European chamois for substrate remains unchanged by temperature [9, 10]. A similar result is noted in kinetic studies of lactate dehydrogenase from skeletal muscle from fishes [43]. As a corollary, the effect of temperature on catalytic rate is equal over the entire (physiological) temperature range. While these conclusions do not allow one to advance the argument that all enzymes are equally affected by temperature (clearly they are not), they do compel one to consider enzyme kinetics, and especially mechanisms of enzyme regulation under various physiologically relevant conditions.

There appear to be three ways in which an organism could maintain the integration of its metabolism over a wide thermal range: (1) the concentrations of intracellular enzymes could be varied in a manner compensatory to thermal change; (2) enzymes with different degrees of catalytic efficiency could be synthesized in response to a chronic or repeated change in environmental temperature, and (3) the intracellular milieu may be altered in such a manner as to render metabolic regulation stable over a certain physiological range of temperatures. In the locally heterothermous mammal and the mammalian hibernator all three mechanisms appear to be operative.

RESULTS AND DISCUSSION

Enzymes from heterothermous tissues: The ears of the varying hare (*Lepus americanus*) are as exposed to environmental temperatures as they are to the observer. The ears are sparsely covered with short hairs, but well endowed with a circulation which prevents them from freezing at the lowest temperatures experienced in the wild. When acclimated to low temperature (-40°C) in the laboratory, variants of several enzymes of glycolysis are found in the exposed ear (Table 1), without significant change in activity.

Hares captured in the wild, where large temperature fluctuations are experienced continually (-35°C to -1°C for the season), show an identical shift in these enzyme variants. However, accompanying this change is a marked increase in the activities of all the enzymes of glycolysis. (To be sure, the diets of the hares described in the present data were clearly different: The wild animals ate natural foods while the laboratory animals ate lab chow.) However, these data prompt the question: Does an animal adapt to cold per se or does it adapt to change in temperature? In the latter case, the hares would have been stimulated more intensely and frequently even though they experienced only once the low temperatures to which the laboratory hares were exposed. Whatever the case may be, the results agree with the common observation that the cold-adapted mammal has a raised metabolic rate, and supports the careful work

TABLE 1

ACTIVITIES OF VARIOUS ENZYMES FROM THE EAR OF THE VARYING HARE

Enzyme	Enzyme Activities Acclimated to -30°C (Lab)	Winter (Wild)	Isoelectric Points Summer	Winter
Phosphofructokinase	6.9 + 1.7	10.3 + 3.4	5.3	5.9
Fructose bisphosphatase	2.3 + 0.6	6.4 + 1.1	7.3	6.9
Aldolase	4.7 + 1.3	9.6 + 0.7	4.6	5.2
Triosephosphate isomerase	935 + 168	1134 + 63	6.4	5.7
Glyceraldehyde-3-phosphate isomerase	4.8 + 2.4	7.1 + 0.4	4.9	4.6
Enolase	13.1 + 4.3	33.4 + 5.2	6.3	6.3
Pyruvate kinase	34.8 + 12.4	63.1 + 21.9	6.1	5.4
Lactate dehydrogenase	67.9 + 14.1	77.9 + 24.7	6.8	6.0

Six animals were acclimated to -30°C, and six were taken from the wild in mid-January. All enzyme activities were determined at 25°C and are expressed as μmole of substrate converted/min./gram tissue. Values of the isoelectric points (pI) of the enzymes from summer and winter hares are listed. Note that the pI values for the -30°C animals and the winter hares are identical (see text); thus only one pI value is listed for the winter animals.

of Joyce [23] who showed that such adaptations may be local, as measured on the rabbit's ear.

Enzyme studies on the hibernator: In the Arctic ground squirrel (*Citellus undulatus*), there appear to be no local thermal regimes, and adaptations observed at the enzyme level appear equally diverse. The Arctic ground squirrel may hibernate for as much as eight to nine months of the year. During hibernation, the animal's metabolic rate may fall to as low as 1/200th of normothermic levels; body temperature falls accordingly, approaching the temperature of the hibernaculum [20]. The animal does not feed during its rather regular and predictable bouts of arousal. These bouts are short but expensive in energetic terms. Fat is by far the predominant fuel of metabolism, although there appears to be a substantial breakdown of body protein.

Over the course of the winter, the gut involutes and the mass of the body musculature decreases appreciably. The major fate of this protein, as well as glycerol from triglyceride fat, is gluconeogenesis, which occurs during hibernation [15], although concentrations of blood glucose vary markedly between bouts of arousal. Indeed, it appears that gluconeogenesis is favored at low temperature in the hibernating ground squirrel [15, 24]. This finding is supported by observation of enzyme variants of fructose 1,6 bisphosphatase (FBPase) in the liver of the Arctic ground squirrel [5].

The winter variants of FBPase retain the salient regulation properties of the summer FBPase, but differ in the temperature range over which regulation could be expressed in a physiologically meaningful way. Further, a decrease in temperature causes an enhancement of enzyme substrate affinity in the hibernating (H) FBPase variants. Hepatic pyruvate kinase (PyK) from the Arctic ground squirrel goes through such an alteration in its variants. The key features of these studies were that PyK occurred as one variant in the nonhibernating (NH) animal and another, with a different isoelectric point, in the hibernating ground squirrel.

Extensive kinetic analysis of the NH-PyK showed that affinity of the enzyme for its substrates and modulators was affected in different ways by temperature. It was almost identical to that of ectothermic enzymes examined over a wide temperature range but at constant pH. In marked contrast, affinity of the H-PyK variant for its ligands was almost completely unaffected by temperature, and constant $S_{0.5}$ values of the substrate and modulators resulted. These experiments were under conditions of constant pH [5].

In this context, it is significant that the pH of extra -and intracellular fluids in the hibernating mammal remain constant over a wide temperature range, thereby rendering these fluids more acidic at the lower temperatures [27, 28]. The hibernator apparently does this by accumulating CO_2, storing it as bicarbonate ion, and then once again giving it off as CO_2 during arousal. The function of this lowered pH during hibernation is unclear at present. It appears, not illogically, that the regulatory characteristics of at least this key enzyme are most clearly observed under experimental conditions that approach the physiological state [7].

The variants of PyK and FBPase in the liver of the Arctic ground squirrel appear in what may be viewed as a circannual cycle. Such cycles may, in part, provide the basis for seasonal thermal adaptation in nature, a view supported by numerous studies mostly of fishes [18, 19, 30, 41].

A difficulty with most studies using ectothermic material is that either the genetics of the isoenzyme systems under study were not known, or the samples utilized were not large enough to expose possible polymorphisms [38]. Although such isoenzyme changes do occur, they appear to be rather rare. It has been suggested that they be confined to organisms, in this case fishes, with a polyploid history [18, 19, 38, 39].

These difficulties were addressed in a large and careful study in isoenzyme patterns in the green sunfish (*Lepomis cyanellus*) [38] in which 12 enzymes from five different tissues were examined during acclimation to a change in temperature and dissolved oxygen. In that study no large changes in isoenzyme patterns were observed. However, significant changes in the activities of numerous enzymes occurred. In our opinion, the most notable feature was that enzymes in a single metabolic pathway, e.g., glycolysis, usually exhibited parallel changes in a given tissue. On the other hand, enzymes in different pathways within the same tissue frequently changed in opposite directions, one group of enzymes increasing while another decreased.

The hibernator is well suited to the study of metabolic reorganization. In winter, during deep hibernation, the animal's heart and lungs receive a major portion of the circulating blood, with a substantial portion supplying the liver and brain. Similarly, brown adipose tissue is active in the winter animal, although it is greatly reduced in mass and activity in the summer-active animal. Some tissues and pathways are active during the summer while others are favored in the hibernating state.

Analysis of enzymes of the pentose shunt, glycolysis and gluconeogenesis in liver, kidney cortex, heart, cerebral cortex and skeletal muscle of the Arctic ground squirrel in both H and NH states yields results qualitatively similar to those of Shaklee et al. [38]. The broad features are an elevation in the activities of the glycolytic and shunt enzymes in both brain and heart muscle during hibernation, a decrease of these enzyme activities in skeletal muscle and liver, while in kidney cortex they are only slightly decreased (Table 2). It is noteworthy that the enzyme essential for hepatic gluconeogenesis, phosphoenolpyruvate carboxykinase, also is elevated in the deep hibernating state. This is in accord with the finding that gluconeogenesis is indeed favored during hibernation [15, 24]. This view is also substantiated in the investigations of two species of hamsters, *Mesocricetus auratus* and *M. brandti*, in which blood glucose levels are regulated during hibernation (Musacchia and Deavers, this volume). To be sure, there are some exceptions to the broad changes outlined above, but in the main one may regard such changes in profiles of enzyme activity as indicative of changes in the overall capacity of the pathways in these tissues.

Are such changes in enzyme activity related to a changeover in enzyme variants, as observed in hepatic PyK and FBPase, or are they the result of either degradation or (reversible) inactivation? Electrofocusing of hepatic 6-phosphogluconate dehydrogenase (6-PG deh) of the pentose shunt reveals only one variant of the enzyme the year round. This variant has a pI of 5.15, although

TABLE 2

RATIOS OF ACTIVITIES OF ENZYMES OF GLYCOLYSIS FROM VARIOUS TISSUES OF THE
HIBERNATING (WINTER) AND NONHIBERNATING (SUMMER-ACTIVE) ARCTIC GROUND SQUIRREL

Enzyme	Cerebral Cortex	Heart Muscle	Kidney Cortex	Liver	Skeletal Muscle (Gastrocnemius)
Hexokinase	271	107	94	35	45
Phospho-glucomutase	138	108	91	52	143
Phosphoglucose isomerase	109	139	98	77	80
Phospho-fructokinase	60	178	114	101	203
Fructose bisphosphatase	-no activity-		46	73	27
Aldolase	136	250	111	76	62
Triosephosphate isomerase	50	54	49	56	53
Glyceraldehyde-3-phosphate isomerase	49	48	43	19	60
Phosphoglycerate kinase	165	148	76	103	35
Phosphoglycerate mutase	144	96	92	119	41
Enolase	119	148	78	84	57
Pyruvate kinase	217	250	76	273	303
Lactate dehydrogenase	132	113	128	279	100

All determinations were carried out at 25°C, and under conditions of optimal
activity. The ratios (activity of hibernating tissue)/(activity of nonhibernating tissue) are expressed as percentages.

in the winter activity of the enzyme is about 30% lower than in summer [40].
Kinetic analysis of the enzyme shows values of the $S_{0.5}$ for either 6-
phosphogluconate or NADP to be unaffected by a change in temperature. As a
result, the Arrhenius plot for the 6-PG deh reaction is linear over
the temperature range tested (5°-37°C) with a calculated energy of activation of
about 59 KJ/mole. The pentose shunt is largely a biosynthetic pathway,
supplying reducing power for lipogenesis.

These findings on the 6-PG deh are, at first glance, at variance with our
previous findings on the liver FBPase from the ground squirrel [5]. However,
while gluconeogenesis must be retained over the entire year, lipogenesis is

active only during periods of growth, membrane synthesis, and, in the case of the hibernator, fattening. In the hibernating ground squirrel, it is not surprising to find a diminished activity of 6-PG deh, which is evenly affected by thermal change. This conclusion is amply supported by comprehensive biophysical and kinetic experiments on purified pyruvate kinase from skeletal muscle of the ground squirrel [32]. As in the case of liver 6-PG deh, the muscle PyK occurs as one variant during the entire year (pI = 6.9). It is noteworthy that activity of the enzyme is enhanced in the hibernating state. The $S_{0.5}$ values of the muscle PyK for substrates and modulators are completely unaffected by thermal change. As a result the Q_{10} of the PyK reaction is constant over the entire physiological substrate and temperature range [32]. This constancy of the kinetic characteristics is supported by a seeming rigidity of the enzyme's conformation with respect to temperature, as examined through the derivative spectra of the purified enzyme under a variety of states [32]. However, the enzyme retains its conformational flexibility in responding to its modulators equally well over the entire temperature range, dispelling the notion that inflexibility in the face of thermal change implies a brittleness of structure which often causes breaks in an enzyme's Arrhenius plots. A detailed discussion is presented in the recent review by Charnock [13].

To test whether the permanence of enzyme variants in a tissue is related to seasonal activity or inactivity of a tissue, we examined an enzyme from brown adipose tissue (BAT) of the hoary marmot (*Marmota caligata*) [6]. BAT increases markedly before hibernation [1]. Although the bulk of the heat of BAT is produced through a variable uncoupling of oxidative phosphorylation [12], which has yet to be demonstrated in the intact cell, there appear to exist a number of cytosolic reactions which could contribute to heat produced by the BAT cell. A logical manner in which the BAT cell could lose energy as heat is through the operation of "futile cycles," which have been described as occurring at the FBPase-PFK and phosphoenolpyruvate carboxykinase (PEPCK)-PyK axes [34]. It has been generally accepted that activity of FBPase is closely regulated, as the simultaneous and rapid operation of both FBPase and PFK within a single cell which would lead to a net hydrolysis of ATP and what has been called a "short circuit in metabolism" [21]. However, the current view holds that control of either enzyme at this plane of regulation may not be complete. The resultant "futile cycle" may serve to enhance flexibility and scope of regulation [34]. In this connection FBPase and PEPCK have been identified in BAT of the fetal and newborn rat [14, 37]. As the enzymes are generally viewed as having a biosynthetic function, these investigators imply that FBPase and PEPCK may play a thermogenic role in BAT of the newborn mammal.

FBPase from BAT of the hoary marmot is found in only one form with an isoelectric point at pH 8.2. Activity of the enzyme is enhanced about sixfold in the hibernating state. This increase in FBPase activity is paralleled by a rise of similar magnitude in PFK activity, suggesting strongly that there is a constant proportionality in the activities of these two key enzymes [25]. The elevated activities of these two enzymes in the hibernating state could increase the amount of heat generated by cycling at this step.

As we observed in the ground squirrel PyK from muscle and G-PG deh from liver, the $S_{0.5}$ values of FBPase for substrate FBP, activator phosphoenolpyruvate (PEP) and inhibitor AMP are unchanged by temperature. The significance of this constancy of the kinetic characteristics is underscored by equally constant concentrations of the metabolites F-6-P, FBP, ATP, ADP, AMP, and PEP [6].

The findings on muscle PyK, liver 6-PG deh and BAT FBPase do not fit neatly into the previous picture of thermal adaptation in the hibernator [5, 11, 32] and in ectotherms in general [19, 33, 41]. In view of accumulating data on the physiology of hibernation, this is perhaps to be expected. Therefore, it should not be surprising to find that not all enzymes, particularly the regulatory ones, adapt in the same way for all tissues from season to season. One may safely accept that brown fat does not have a thermogenic role in the adult non-hibernating subject because total mass of the tissue and enzyme activity are much reduced in the summer-active state [1]. A similar argument is made on the basis of the data on hepatic 6-PG deh and muscle PyK, enzymes from pathways which are much reduced during hibernation. There would be no adaptive value in elaborating seasonal variants of some enzymes from such pathways.

SUMMARY AND CONCLUSIONS

The evolution of hibernation must necessarily have involved the incorporation of mechanisms which facilitate adaptation to temperature at the enzyme level, mechanisms which function over a time course of seconds and throughout the seasons. From available physiological data it is clear that while the organism adapts as a whole, not all tissues and pathways must adapt in the same way and clearly they cannot. This is readily seen from such examples as the profound changes in circulatory patterns and extended starvation during hibernation. This would make good sense, metabolically. The elaboration of enzyme variants in all pathways or tissues, or the (reversible) activation-inactivation of the many enzymes possible would impose a large energetic drain on the animal. The observed changeover in enzyme variants seems to be limited to parts of pathways which are required in both the homeothermic (summer-active) and the hibernating states. The remaining metabolic sequences, and here one may include identical pathways but in other tissues, do not undergo a changeover in enzyme variants, thus enabling the organism to maintain a strict metabolic economy.

In every system of which we are aware, the regulatory properties of a metabolic reaction or sequence are best expressed under conditions that approach as closely as possible the physiologic state. While this approach is often more laborious and complex for the investigator, it should restrict erroneous conclusions regarding data that superficially seem to be reflections of an adaptive strategy [5, 19, 31]. Chief features in this context are the related factors of intracellular pH and temperature. Further, physiological and biochemical studies of the hibernator, *Citellus*, indicate strongly that the animal exists as different entities at different times of the year. For example, BAT in the adult hibernator would seem to have no role to play in the summer-active state; neither would the pentose shunt and lipogenesis in liver, or large glycolytic activity in skeletal muscle during winter. Metabolic reorganization does occur, but it does so in a complex pattern in which certain tissues and metabolic sequences take temporary precedence, and has evolved in such a way as to minimize the loss of energy.

It seems imperative to us to gain insights at four different levels of organization. First, on the level of the whole animal, such as the hare, does it adapt to low temperature per se, or is the stimulus for adaptation the thermal change itself? The integration of metabolic responses in differently adapting tissues would seem an especially fertile area of research. Further, what are the physiological triggers for such localized adaptive responses?

Second, the observed change in the enzyme variant profile in some tissues, e.g., liver FBPase and PyK, must be integrated at a higher level, and also at the genetic level of regulation. It is of extreme interest if the appearance of new variants is regulated at the transcriptional or the posttranscriptional plane. This question would seem especially fruitful for studies on the putative stimulus that is responsible for the preparation and onset of hibernation.

Third, elegant work has been done in recent years on the relationship between temperature and acid-base regulation cells. In our view, many of the previous explanations of adaptive strategies have been seriously questioned [11, 18, 19, 31, 41], including some of our own [4, 5, 6, 8]. Much work is needed in this area, especially in view of the interesting observation by Malan [27, 28] that intracellular pH is not equally effected by temperature in the various tissues of the hibernator. Especially perplexing is the role of enhanced acidity of the hibernator's blood and intracellular fluid of various tissues, except liver, during hibernation.

Fourth, at the enzyme level, questions arise on the regulation of enzyme activity. The interaction between enzyme turnover and metabolic interconversion between active and inactive forms are items of special interest and promise in this regard. Related to this is the study of what is altered in the molecule during the synthesis of appearance of new enzyme variants. The application of new analytical techniques which probe enzyme conformation and amino acid composition [32, 33] would seem to be particularly useful in this regard.

In sum, animals which in regular and predictable fashion preadapt to seasonal variations in the environment are satisfying experimental tools. Using such organisms, it is possible to dissect the relevant factors from the artifacts of the experiment. It is difficult to speculate from the laboratory to the natural state where animals continue to develop their habits. Irving [22], proposing one of the more enlightened concepts, said "in spite of recognizing the difficulty of projecting ideas from here to there and from now to then, speculation is still the most rapid way to enlarge one's thoughts."

ACKNOWLEDGMENT

The experiments described in this paper were supported by a grant from the National Institutes of Health (NIGMS 10402).

REFERENCES

1. Afzelius, B. A. (1970) Brown adipose tissue: Its gross anatomy, histology and cytology, in Brown Adipose Tissue, Lindberg, O., ed. American Elsevier, New York, pp. 1-31.

2. Atkinson, D. E. (1977) Cellular Energy Metabolism and Its Regulation. Academic Press, New York.

3. Baldwin, J. and Hochachka, P. W. (1970) Functional significance of isoenzymes in thermal acclimatization: Acetylcholinesterase from trout brain. Biochem. J., 116, 883-887.

4. Behrisch, H. W. (1969) Temperature and the regulation of enzyme activity in poikilotherms. Fructose diphosphatase from liver of migrating salmon. Biochem. J., 115, 687-696.

5. Behrisch, H. W. (1978) Metabolic economy at the biochemical level: The hibernator, in Strategies in Cold: Natural Torpidity and Thermogenesis. Wang, L. C. H. and Hudson, J. W., eds. Academic Press, New York, pp. 461-497.

6. Behrisch, H. W. (In press) Temperature and regulation of enzyme activity in the hibernator. Fructose diphosphatase from brown adipose tissue of the hoary marmot *Marmota caligata*. Can. J. Zool.

7. Behrisch, H. W. and Galster, W. A. (In press) Temperature and the regulation of enzyme activity in the hibernator: Effects of cations on pyruvate kinase from liver of the Arctic ground squirrel. J. Thermal Biol.

8. Behrisch, H. W. and Hochachka, P. W. (1969) Temperature and the regulation of enzyme activity in poikilotherms. Properties of rainbow trout liver fructose diphosphatase. Biochem. J., 111, 287-295.

9. Behrisch, H. W., Ortner, I. and Wieser, W. (1977) Temperature and the regulation of enzyme activity in heterothermic tissues of an alpine mammal. Lactate dehydrogenase from skeletal muscle of the chamois. J. Thermal. Biol., 2, 185-189.

10. Behrisch, H. W. and Percy, J. A. (1974) Temperature and regulation of enzyme activity in homeothermic and heterothermic tissues of arctic marine mammals. Some regulatory properties of 6-phosphogluconate dehydrogenase from adipose tissue of the spotted seal *Phoca vitulina*. Comp. Biochem. Physiol., 47B, 437-443.

11. Borgmann, A. I. and Moon, T. W. (1976) Enzymes of the normothermic and hibernating bat *Myotis lucifugus*. J. Comp. Physiol., 107, 185-199.

12. Cannon, B., Nedergaard, J. and Sundin, U. (1981) Thermogenesis, brown fat and thermogenin, in Survival in Cold, Jansky, L. and Musacchia, X. J., eds. American Elsevier, New York.

13. Charnock, J. S. (1978) Membrane lipid phase-transitions: A possible response to hibernation?, in Strategies in Cold: Natural Torpidity and Thermogenesis. Wang, L. C. H. and Hudson, J. W., eds. Academic Press, New York, pp. 417-460.

14. Frohlich, J., Hahn, P., Kirby, L. and Webber, W. (1976) Rat fetal brown fat in vitro: Effects of hormones and ambient temperature. Biol. Neonate, 30, 40-48.

15. Galster, W. A. and Morrison, P. R. (1970) Cyclic changes in carbohydrate concentrations during hibernation in the Arctic ground squirrel. Am. J. Physiol., 218, 1228-1232.

16. Hazel, J. and Prosser, C. L. (1974) Molecular mechanisms of temperature compensation in poikilotherms. Physiol. Rev., 54, 620-677.

17. Helmreich, E. and Cori, C. (1964) The effects of pH and temperature on the kinetics of the phosphorylase reaction. Proc. Natl. Acad. Sci. US, 52, 647-654.

18. Hochachka, P. W. and Somero, G. N. (1971) Biochemical adaptation to the environment, in Fish Physiology, Hoar, W. S. and Randall, D. J., eds. Academic Press, New York, vol. 6, pp. 99-156.

19. Hochachka, P. W. and Somero, G. N. (1973) Strategies of Biochemical Adaptation. Saunders, New York, 358 pp.

20. Hock, R. (1960) Seasonal variations in physiologic functions in Arctic ground squirrels and brown bears, in Mammalian Hibernation, Lyman, C. P. and Dawe, A. R., eds. Bull. Museum Comp. Zool. (Harvard), 124, 155-171.

21. Horecker, B., Pontremolis, S., Rosen, O. and Rosen, S. (1966) Structure and function in fructose diphosphatase. Fed. Proc., 25, 1521-1528.

22. Irving, L. (1972) Arctic Life of Birds and Mammals. Springer-Verlag, Berlin, Heidelberg, New York.

23. Joyce, C. M. (1969) The effect of environmental temperature on succinic dehydrogenase activity in the ear skin of the rabbit. Proc. Royal Irish Acad., 65B, 425-433.

24. Klain, G. J. and Whitten, B. L. (1968) Carbon dioxide fixation during hibernation and arousal from hibernation. Comp. Biochem. Physiol., 25, 363-366.

25. Klingenberg, M. and Pette, D. (1969) Proportions of mitochondrial enzymes and pyridine nucleotides. Biochem. Biophys. Res. Comm., 7, 430-432.

26. Malan, A., Duall, F. and Rodeau, J. L. (1976) Relations entre les etats acide-base extra- et intracellulaires dans l'acidose de l'hibernation. J. Physiol. (Paris), 72, 105A.

27. Malan, A., Warens, H. and Waechter, A. (1973) Pulmonary respiration and acid-base state in hibernating marmots and hamsters. Respir. Physiol., 17, 45-61.

28. Malan, A., Wilson, T. and Reeves, R. B. (1976) Intracellular pH in cold-blooded vertebrates as a function of body temperature. Respir. Physiol., 28, 29-47.

29. Miller, L. K. (1976) Production of threitol and sorbitol by an adult insect. Association with freezing tolerance. Nature, 258, 519-520.

30. Moon, T. W. (1975) Temperature adaptation: Isozymic function and maintenance of heterogeneity, in Isozymes II, Physiological Function, Markert, C. L., ed. Academic Press, New York, pp. 207-220.

31. Moon, T. W. and Borgmann, A. I. (1976) Enzymes of the normothermic and hibernating bat *Myotis lucifugus*. Metabolites as modulators of pyruvate kinase. J. Comp. Physiol., 107, 201-210.

32. Morse, G. and Behrisch, M. W. (In press) Temperature and the regulation of enzyme activity in the hibernator. A kinetic and spectroscopic study of pyruvate kinase from skeletal muscle of the Arctic ground squirrel. Can. J. Biochem.

33. Morse, G. and Behrisch, H. W. (In press) On the application of derivative spectroscopy in differentiating between isoproteins. Anal. Biochem.

34. Newsholme, E. A. and Start, C. (1973) Regulation in metabolism. Wiley, New York.

35. Reeves, R. B. (1969) Role of body temperature in determining the acid-base state in vertebrates. Fed. Proc., 28, 1204-1208.

36. Reeves, R. B. (1977) The interaction of body temperature and acid-base tolerance in ectothermic vertebrates. Ann. Rev. Physiol., 39, 559-586.

37. Seccombe, D. W., Hahn, P. and Skala J. P. (1977) Fructose-1, 6-diphosphatase activity in brown adipose tissue of the developing rat. Can. J. Biochem., 55, 924-927.

38. Shaklee, J. B., Christiansen, J., Sidell, B., Prosser, C. L. and Whitt, G. S. (1977) Molecular aspects of temperature acclimation in fish. Contribution of changes in enzyme activities in isoenzyme patterns to metabolic reorganization in the green sunfish. J. Exp. Zool., 201, 1-20.

39. Sidell, B., Wilson, T., Hazel, J. and Prosser, C. L. (1973) Time course of thermal acclimation in goldfish. J. Comp. Physiol., 84, 119-127.

40. Smullin, D. H. (1980) Temperature and the regulation of enzyme activity in a hibernator, 6-phosphogluconate dehydrogenase from liver of the Arctic ground squirrel. University of Alaska, M.S. Thesis.

41. Somero, G. N. (1975) The roles of isozymes in adaptation to varying temperatures, in Isozymes II, Physiological Function, Markert, C. L., ed. Academic Press, New York, pp. 221-234.

42. Taketa, K. and Pogell, B. (1965) Allosteric inhibition of rat liver fructose 1,6-diphosphatase by adenosine 5'-monophosphate. J. Biol. Chem., 240, 651-662.

43. Wilson, T. L. (1977) Interrelation between pH and body temperature for the catalytic rate of the M4 isozyme of lactate dehydrogenase (EC 1.1.1.27) from goldfish *Carassius auratus*. Arch. Biochem. Biophys., 179, 378-390.

44. Wilson, T. L. (1977) Theoretical analysis of the effects of two pH regulation patterns on the temperature sensitivities of biological systems in nonhomeothermic animals. Arch. Biochem. Biophys., 182, 409-419.

Author Index

Subject Index